Life in Action

Life in Action

Action

biochemistry explained

Peter Farago
and John Lagnado

VINTAGE BOOKS

A DIVISION OF RANDOM HOUSE
NEW YORK

FIRST VINTAGE BOOKS EDITION, March 1973

Copyright © 1972 by Peter Farago and John Lagnado

Library of Congress Cataloging in Publication Data

Farago, Peter, 1932–
  Life in action.
  Reprint of the 1972 ed.
  1. Biological chemistry. I. Lagnado, John,
1933–   joint author. II. Title.
[QP514.2.F37 1973]   574.1'92   72–8706
ISBN 0–394–71906–9

Manufactured in the United States of America

Preface

AMONG the life sciences biochemistry plays a central role deciphering and illuminating the functions of living organisms. It has evolved a system of ideas and techniques which are successfully applied all the way from industrial processes to the complex hardware of the laboratory. The work of living cells, from the fermentation of sugars to the transmission of messages in the brain, is part and parcel of biochemical studies.

Even more important and interesting, is the influence of biochemical ideas on man's view of himself: not surprisingly perhaps a number of biochemical practitioners have also taken active roles in considering the place in society of man and scientist.

In this book we have attempted to describe some of the more important concepts of biochemistry, the experiments on which they came to be based, and the implications they have for the future. But *Life in Action* is not a textbook. It is primarily intended for the interested non-scientist with little or no training in biology and chemistry. Nevertheless we hope that those trained in the 'harder' sciences might also derive benefit and enjoyment from seeing how their ideas can be used in a biochemical context.

In many areas of biochemistry the pace of advances continues at an explosive rate. Since this book has gone to press, Khorana and his team has synthesised genes, and controversy over the central dogma of replication has reached new heights. New emphasis on the problem of ecology and environment are bringing to the fore biochemical considerations: the use and abuse of enzymic washing powders, insecticides, drugs, and nerve gases have all passed from the learned journals into everyday consciousness. We hope that *Life in Action* will help in enabling the non-biochemist to place new developments in perspective and weigh up the implications of biochemical discoveries.

We thank Professor S. P. Datta, Professor L. Young and Dr Keith Howlett for their constructive criticisms; Mr Harold Strauss for his editorial labours, Sheila Watson, of Bolt and Watson, for her constant encouragement; and Jenny Lagnado for her help with the illustrations. The photographs for Figure 17 were kindly provided by Dr J. C. Kendrew. All errors of omission or commission are ours.

1972   P.J.F. and J.R.L.

Contents

Life in Action

CHAPTER 1

biochemistry: its purpose and scope

Biochemistry is an attempt to offer a consistent set of explanations for all processes occurring in living matter in terms of chemical events. The nature of these processes has offered a fascination and a challenge to man's intellect from the very dawn of civilization; it has inspired not only scientific theory but also poetic expression and religious thought. Biochemistry has come on the scene relatively late; it is still a young and sometimes aggressive discipline. The greatest and most important biochemical advances have virtually all been made in the last twenty years; the pace of research is increasing all the time and, most pleasingly, biochemistry is still resounding from disputes among its practitioners.

But what exactly is biochemistry? It is an inquiry into the chemical processes that make life possible in all animals and plants, through an investigation into the nature of their materials and their reactions which provide the large-scale manifestations of life. In order to carry out the necessary observations and interpret the data thus obtained, biochemistry uses the tools, data, and ideas derived from other disciplines and adapted to its own needs. Biochemistry has become the marketplace where the older sciences exhibit their offerings and allow the biochemist to choose the most suitable: the structures, locations, and functions of organs from anatomy and physiology; the knowledge of chemical substances from chemistry; the methods of X-ray crystallography, ultracentrifugation, and electrical conduction from physics.

If this description seems too forbidding, we can find an analogy between biochemistry and the sciences allied to it by considering what would happen if we applied the same notions to the study of motorcars. The principle on which the works of all cars is based is chemistry: it concerns statements about the combustion of petrol in air; this we can compare to biochemistry. The description of where all the components of a car are located is equivalent to anatomy; the knowledge of how such parts as pistons or steering mechanisms work is dealt with in physiology. Testing of the components is a matter for physics. The choice of the correct oils and fuels is the province of the nutritional scientist, and, after a crash, the physicians and surgeons take over.

Biochemistry takes the known facts of chemistry and applies them to the living organism. That it can do this depends on a notion that there is nothing fundamentally different between chemical reactions taking place in a test tube and those taking place in a bacterium or a human being; in consequence, all chemical compounds made by any organism can be made in the laboratory, at least in theory. This idea is the complete negation of what used to be called the life-force theory, which assumed that although living organisms could be regarded as consisting of a vast complex of chemical reactions, life itself contributed some additional, intangible quality.

The controversy of vitalists versus mechanists went on for the best part of seventy years with practically everyone in the world of science joining in, including Pasteur and Liebig. One of the critical issues in this argument centred around the question whether a process such as fermentation (sugar turning to alcohol) needed living yeast cells in order to take place, or whether it could be described in purely mechanistic terms. In 1896, Buchner tried using yeast juice, without the presence of cells, to ferment sugar to alcohol. He succeeded.

The life-force theory was initially challenged in 1828, when a chemist named Wöhler prepared urea, a constituent of urine, in the laboratory, using conventional bench chemicals. Not all Wöhler's fellow scientists were convinced; the very eminent chemist Berzelius suggested that Wöhler might end up in the same liquid his beloved urea had come from.

Biochemistry pursues the lowest common denominator of all

forms of life. This denominator is the *cell*, and it is interesting to note how the word 'cell' came to be used in its present sense. In 1665 Robert Hooke, of Hooke's Law fame, was looking at slices of cork through a microscope. He was intrigued to see what he described as a collection of small box-like rooms, like those inhabited by the religious orders. Not unreasonably, he called these structures cells.

What bricks are to houses, cells are to all living organisms. They are the units of life, on which all other functions depend. What we are or can do depends in the last analysis on the arrangements and types of cells we possess—the average human being has about five million million of them. Biochemistry is concerned with chemical processes taking place in cells, and, as we shall see, the basis of all cell behaviour is a multitude of interconnected chemical reactions. We can turn this idea around and say that a cell can be described in terms of the chemical reactions it is capable of performing.

All living organisms are made up from conglomerations of cells, called *tissues*, and looking at the astounding variety of life on this planet, we would expect to find a completely baffling variety of cells. Luckily for us, this is not quite so. Although there are numerous types of cells, differing according to where they are and what they have to do, similarities among cells are far greater than differences. A yeast cell and a human muscle cell are still recognizably of the same ilk: the fundamental biochemical differences between man and bacteria are much smaller than we would perhaps like to believe. In a roundabout way, biochemistry agrees with theology in stating the fundamental unity of all living things.

What do we gain by learning about the chemistry of these units of life? There are two answers to this question.

On the level of expediency, a knowledge of biochemical processes is certainly very useful. Wherever food is prepared or eaten, biochemical processes take place. The fast development of medicine during the last fifty years, the discoveries and improvements of medicines of all types from vaccines to antibiotics, owe a great deal to the biochemical knowledge that not only helped along their development but also created the possibilities for their manufacture and large-scale exploitation. As our understanding of the processes taking place in plants and animals improves, we shall be in a far better position to cure disease or, if possible, prevent it.

In industry too, biochemistry is contributing increasingly to better and safer products. To the traditional industries based on biochemistry such as the brewing of beer, fermenting of wine, tanning of leather, producing of cheese, are added new ways of using biochemistry. The knowledge of fermentation processes not only allows us to make antibiotics, it can also be used in the processing of petrochemicals, for instance, in removing waxy fractions that are difficult to manipulate, and it offers the prospect of producing enough synthetic foods from oil or waste products to make a significant contribution to the problem of feeding the increasing world population.

Biochemistry also makes a contribution to the understanding of our own thought and behaviour patterns. Learning and memory, according to our present ideas, may be founded on the existence of biochemical functions. Drugs can modify human behaviour and have already revolutionized the treatment of certain mental diseases. They have also furthered our understanding of how the human brain functions.

The second argument for the study of biochemistry is simply man's love of knowledge. We like to know what makes us tick, we like to respond to the challenge of the unknown both within and around us. We have already learned a lot, but biochemistry, still in its infancy, will undoubtedly give us a far better insight into all forms of life.

In trying to understand the thought processes underlying biochemical work we must also be aware that we are dealing with matters of great actual or potential power. The results of biochemistry can be abused as well as used. They give the promise of a better future for mankind but may also threaten our very existence. The great controversy over contraceptive drugs has shown that biochemistry, by allowing us to manipulate our most fundamental functions, can pose questions that cannot easily be answered. Chemical and biological warfare agents raise problems that are essentially those of public morality, but they have arisen through increasing biochemical knowledge. We do not have to believe the dreams or nightmares of science fiction or sensational science journalism to realize that biochemistry offers not only a scientific but also a moral challenge.

TECHNIQUES USED IN BIOCHEMISTRY

We have mentioned earlier that the unit of all life is the cell, the building material of all plant and animal tissue. Because various parts of the body have to fulfil different functions, they are built from different types of cells: red blood corpuscles and white ones, liver cells and bone cells, plant cells and nerve cells are all different. But they have enough features in common for us to be able to talk, at least to start with, about a generalized cell.

In order to see such a cell and describe its contents, biochemistry had to wait for other sciences and technologies to provide it with the necessary experimental tools. In Hooke's time the microscope was already pressed into service and, as a result of continuous developments, finer and finer details of the cell could be distinguished.

The limit of microscopy was reached when biochemists wanted to look at structures smaller than the wavelength of the light used. This cannot be done. However, further developments led to the electron microscope which uses beams of electrons instead of waves of light. It is possible to think of these electron beams as light of a very short wavelength, and the shorter the wavelength of light the smaller the object it is possible to see. Using an electron microscope, objects several hundred times smaller than those visible under a light microscope can be seen, down to the structures of very large molecules.

Another technique was also developed to probe the fine structures of molecules. This was X-ray crystallography, which revealed the structures of various molecules that play very important roles in the cell. Continuous scientific developments therefore allowed the biochemist to see smaller and smaller objects and the biophysicist to probe the molecular architecture of cellular constituents. This process of miniaturization is illustrated in Figure 1.

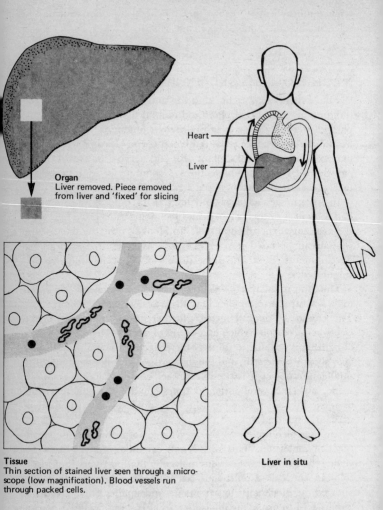

Organ
Liver removed. Piece removed
from liver and 'fixed' for slicing

Heart

Liver

Tissue
Thin section of stained liver seen through a microscope (low magnification). Blood vessels run
through packed cells.

Liver in situ

Figure 1   Organ to Cell

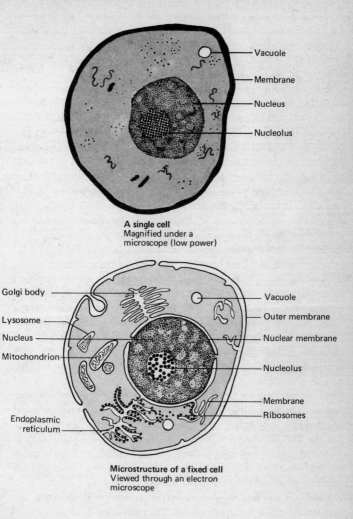

A single cell
Magnified under a
microscope (low power)

Microstructure of a fixed cell
Viewed through an electron
microscope

Figure 1 (continued)

It is not enough to look at a cell; it is also necessary to lift it out of its surroundings and to examine its constituents. Here again, biochemists use an array of complex experimental techniques, the objectives of which are to deliver a given constituent in as pure a form as possible. One of the most important of these methods is differential centrifugation.

Basically, this is the process used in spin-drying clothes. Centrifugal force creates a gravity gradient, and materials of different density line up in order along it. This method has enabled biochemists to break down the constituents of cells into fairly homogeneous groups which can be purified further. In this way, some idea can be obtained not only of the constituent parts of the cell, but also what the function of each of them is. Ultracentrifugation, using a centrifuge at up to 500 000 times normal gravity, is also employed in the purification of some very large molecules.

Yet another technique is chromatography; since it is used a great deal in purifying and manipulating a wide range of biochemically interesting molecules, we shall be dealing with it at greater length in a later chapter. The basis of the technique is that it is possible to separate accurately very small amounts of different materials on the basis of their ability to be adsorbed to different extents on the same surface. Such a surface may be paper, or a resin, or a layer of liquid. The mixtures of substances can be conveyed to the surface in solution, or in a gas stream, and taken off the surface again by some appropriate carrier. In this way, very complex mixtures can be separated into their components, and the nature of these constituents further investigated.

All these methods are essentially the methods of modern analytical chemistry, although X-ray techniques are also part of what is generally considered physical chemistry. In the latter domain fall the various spectroscopic techniques that rely on the absorption and emission of light energy by molecules; in biochemistry these are used essentially as fingerprinting devices to ascertain the actual compound found in a particular experiment.

Finally, there is also a range of experimental techniques that are particular to the living organism. Their differences from classical chemical determinations arise from the fact that, in test-tube chemistry, once a compound is made, it is pretty obvious how it was built up and what one can do with it. Not so in living systems. The fact that a compound has been isolated tells us little; one must also know where it appears in the chemical cycles of the organism. Thus, it may be the precursor or the end product or the middle step of some long and complex series of reactions, and only by knowing its position in the series can its importance be judged. Methods must therefore be available to trace the fate of a compound, or part of a compound, through its travels along the chemical highways of the body. The techniques used to obtain such knowledge rely mainly on tagging atoms: by substituting for a normal atom one of virtually identical chemical properties but of different weight, it is possible to follow the molecule containing it through a number of transformations. The main isotopes used for this work are those of hydrogen, carbon, sulphur, nitrogen, and phosphorus; their ready availability since the last war has helped biochemical research a very great deal.

The complexity of biological systems also means that really good work has to be thorough on a number of counts or else it lacks credibility. It has to be accurate by the criteria of classical chemistry; the chemical compounds it works with have to be reasonably pure and measurements on them must be made so as to satisfy the searching criticisms to which all chemical work is subjected. But in addition, it has also to place its chemical findings within the context of a living organism and show that its explanation is plausible. It is a tribute not only to the technical and scientific skill of its best practitioners but also to their imaginative ability that in a very great number of instances both these criteria are satisfied.

Ideally, the sequence goes something like this. Experiments with the whole animal suggest the existence of a compound and associated reaction systems. These are tracked down to a part of the body which can be isolated. The existence and actions of the substance are further narrowed down by work with thin tissue slices, tissue solutions, and eventually with individual cells or their purified constituents. At the same time, experiments must also begin from the opposite direction in attempting to synthesize the compound or re-create the biological environment in which it operates in the living organism.

Thus the ideal end result is that not only is the compound known by its biological action, but a complete simulated biochemical system is also synthesized in which the functions of the living organism can be reproduced. Although this stage has been reached with relatively few systems up to now, one is near enough to ideal conditions with many others to be reasonably sure of the meaning of evidence already at hand.

THE USE OF INFORMATION THEORY IN BIOCHEMISTRY

Biochemists have a difficult furrow to hoe. Their materials are often fragile and always complex. They have to be isolated, handled, and manipulated carefully, otherwise they refuse to co-operate and are destroyed. Techniques first had to be developed to perform these operations before meaningful measurements could be made.

Measurements themselves had to be adapted to the special needs of biochemistry. Usually this took the form of marrying the developments of physical chemistry—centrifugation, X-ray crystallography, electron microscopy, spectroscopy—to the requirements of biochemical measurements. The specialization most interested in these hybrid methods is called biophysics.

The development of measuring and handling methods is still proceeding at a fast rate. Even today a great deal of the experimental work a biochemist has to do is tedious in the extreme, and to relieve the worker's boredom automatic techniques are now being developed.

The general theory used as a basis of interpreting biochemical events has been derived from classical chemistry. In order that all theories might be suitable for the complex situations of biochemistry, a degree of sophistication had to be reached—in particular it was necessary to become accustomed to handling, and talking about, very large molecules and the effects of a number of chemical reactions going on at the same time.

It is important to realize that despite their difficulties, these complex systems often involved nomenclature deriving from the pre-scientific age of biology, and despite the frequently implied assumption that reactions going on in living organisms are sometimes different from those observable in a test-tube, biochemistry is firmly grounded in the physical sciences and the belief that all fundamental principles are equally valid in all of them. There are differences of scale and differences of complexity, but the underlying principles are the same.

What distinguishes biochemistry from classical chemistry is the complexity of its systems—and one can define systems as being a collection of chemical events, reactions, which all influence each other in some way. Whereas the true chemist generally deals with very small molecules, having molecular masses in the hundreds or less, the biochemist's interest often hinges on enormous entities with molecular masses ranging from thousands to millions, which have a great deal in common with the long atomic chains of the molecules of plastics.

Because the molecules dealt with in biochemistry are frequently very large and because a great many different types of them are engaged in related chemical reactions, the idea has grown up of regarding them not only as chemical raw materials which, with some luck, can be purified and put in a jar, but also as storehouses of information. This idea is especially important in genetics, since characteristics are clearly inherited from one generation to another and it would seem reasonable to assume that an actual physical thing is being handed down from parent to offspring.

One of the most celebrated biochemical researches, on the nature and functions of the nucleic acids (deoxyribonucleic acid, or DNA, and ribonucleic acid, or RNA), has, for example, related our knowledge of chromosomes, the carriers of genetic material, to that of the molecular nature of such material. Current research is also seeking biochemical explanations for the mechanisms of memory, sleep, and the functions of the brain generally.

The idea of information-storing and information-handling comes under the general heading of information theory of cybernetics, and its concepts derive, paradoxically enough, from

Wiener's mathematical work during the last war on how to improve the performance of anti-aircraft batteries. Another early
example of the use of information theory was a mechanical mouse
that would approach a light and not fall off the edge of a table—a
great comfort to those who desired mechanical mice that ap-
proached lights. As it became further developed, information
theory did make, and continues to make, valuable contributions
to activities ranging from telecommunications to factory manage-
ment. In biochemistry also, it has helped research workers to
devise constructive experiments to test hypotheses. One could
speculate on the future possibility of an emerging unity between
biochemistry and information theory: for example, it may become
possible for computers to use biochemical systems as memory
stores.

One effect of using the ideas of information systems is that biochemists sometimes talk in terms of a cell, or parts of a cell, coding,
designing, manufacturing, instructing—a lively and picturesque
way of talk which only becomes dangerous if the notion is put
across that there is some very small man, the brother of the one
inside the television set, who actually performs these actions. There
is, of course, nothing of the sort—and when we say that a cell gives
instructions to manufacture a certain chemical, all we mean is that
chemical reactions took place the end effect of which was the
appearance of a given substance.

Or, as in everyday life, every action has consequences which
somewhere along the line affect the originator of the action and
modify its subsequent behaviour. As in Donne's sermon, the biochemical bell tolls for everyone.

THE CELL

We shall now make a more thorough acquaintance with the cell,
since almost all the rest of this book will be concerned with one
aspect or another of what is happening within cells.

Before starting our description, we must be aware that the cell
we are describing is a generalized, and therefore a somewhat oversimplified, entity which cannot be found anywhere in nature. It is

rather like describing the average Englishman or Frenchman—he does not exist but, we hope, has a sufficient number of attributes shared by all his fellows. For proper characterization we should follow industrial practice and give a job description: what an actual cell does and how. This aspect will be tackled in a further chapter, under the somewhat more scientific title of cell specialization.

As is the nature of all specialized occupations, whether in science, the arts, or interior decorating, a language has grown up in biology and biochemistry for describing the materials and tools used. This makes understanding at a glance more difficult, but on the other hand avoids the necessity of tedious descriptions every time an idea or a physical object is spoken of.

Our ideal cell is shown in Figure 1. Laid side by side, about 1000 of them would measure a centimetre and all together would weigh no more than one ten-thousandth of a gramme. Although our cell is shown as circular, or roughly spherical in reality, this is not always so in practice: shapes of cells vary according to where they are and what they do.

Right at the edge of the cell lies a *membrane*, a sort of biochemical 'iron curtain' put there for much the same reason: to control entry to and exit from the cell. All cells live surrounded by water (it is one of the highlights of biochemical history that Huxley is supposed to have described the Archbishop of Canterbury as consisting of 65 per cent water). So if the cell were not surrounded by an efficient membrane, the surrounding water would leach out all the various substances held inside it and the cell would die.

The membrane also has what are called selective functions. It normally allows only certain types of materials required for correct functioning to enter into the interior of the cell, although some drugs and poisons have the capacity of passing through. Similarly, the membrane acts as a barrier for the cell components, but allows waste products to leave. The actual materials of construction of the cell membrane are fatty materials, called *lipids*, together with *proteins*. Variations do occur according to species; in plants, for example, there is a cell wall outside the cell membrane made of a fibrous material called *cellulose*, for greater rigidity.

Once we pass through the cell membrane, there are two major areas to consider. In the idealized middle of the cell sits the cell *nucleus*, surrounded by a membrane of its own. Under the microscope it appears as a dark, granular body. The nucleus is the headquarters and the control mechanism of the cell—in general, it regulates most of the cell's synthetic work and determines the all-essential division or replication. It is the main storehouse and transmitter of genetic information.

Although the main components of the nucleus are known, there is still a fair amount of controversy and ignorance about their precise relationships and functions. This whole problem is so interesting that we shall be dealing with it at some length in a separate chapter.

In brief, the nucleus contains a set of giant molecules, called DNA's (deoxyribonucleic acids), usually segregated in structures called *nucleoli*. The sets of giant molecules are called *nucleic acids* and are usually found in association with proteins called *histones*. Nucleic acids and histones together are referred to as *nucleoproteins*. In addition, the cell nucleus also contains enzymes.

A feature of the cell is its ability to divide, resulting in the formation of two identical daughter cells. During such division, called *mitosis*, the genetic material contained in the nucleus is handed over in two equal and identical parts to the daughter cells. On a microscopic scale, this material is contained in the *chromosomes*, long rod-like structures. Characteristics of the individual are governed by *genes* which are strung along the chromosomes. On a molecular scale, it is now known that the chromosomes are formed by the nucleoproteins.

Outside the nucleus lies the factory floor of the cell, called the *cytoplasm*. Here the cell's products are manufactured, incoming material is sorted, and waste products are dispatched, all under the general direction of the nucleus. Raw materials required by the cell are sometimes stored in spaces called *vacuoles* within the cytoplasm.

The various components of the cytoplasm can be separated in the centrifuge. The remaining clear fluid contains a variety of simple substances and many enzymes required for the work of the cell. It is best to consider each of the various structures within the cytoplasm under its own heading:

Mitochondria are, on the scale of the cell, fairly large, egg-shaped bodies, surrounded by a smooth outer membrane. Inside the membrane there is an inner one, folded in long zigzags which traverse the whole interior of the unit. The function of the membrane is similar to that of the cell wall, the control of the entry and outflow of material. Within the mitochondrion are arranged in order—the precise order being still in dispute—the enzymes necessary for the energy-supplying reactions of the cell. It has been calculated that up to 75 per cent of all energy required is supplied by reactions taking place in the mitochondria, which one can therefore regard as the powerhouse of the cell.

Lysosomes are small capsules containing a variety of digestive enzymes capable of breaking up large molecules and membrane fragments. They have been called suicide packets, since if all the enzymes were allowed to escape they would be able to destroy the complete cell structure. Lysosomes carry out the necessary scavenging operations if a cell becomes injured.

The endoplasmic reticulum is a sack-like system bounded by a membrane of the same type as the outer cell wall. Its functions are diverse: structural support for the cell, a large surface area for reactions to take place, pathways for the transport of material, and a collection depot for synthesized materials.

There are two types of reticulum: a rough and a smooth. The rough reticulum can be seen in cells engaged in protein synthesis; the rough surfaces are associated with clusters of *ribosomes*, structures where the synthesis of proteins takes place. When tissue is homogenized it is possible to separate out, for instance by differential centrifugation, the fragments of the endoplasmic reticulum together with their attendant clusters of ribosomes. Such fragments are called *microsomes* and are visible under the electron microscope, although they are too small to be seen with an ordinary light microscope.

The smooth endoplasmic reticulum is rich in the fat-like *phospholipids*. It is a membranous structure consisting of proteins called *lipoproteins*. The smooth endoplasmic reticulum is important in the cell's synthesis of lipids and proteins and a number of other chemicals, for instance sex hormones in the testes.

There is a further structure, similar to the smooth endoplasmic reticulum, called a *Golgi complex*. The function of this entity is not clear, although Golgi complexes are probably involved in the fat metabolism of the cell and in secretory functions.

In addition to these major constituents, there are also a host of small entities within the cell; pigments, droplets of oil, and in plant cells a number of special constituents. The functions of some of these are known, but others up to now have defeated all investigations.

With sundry variations, the idealized cell we have looked at exists in all animal and plant life. The smallest living organisms are probably the *viruses*—it is indeed arguable if viruses can be described as living. They possess little of the elaboration of the more sophisticated cells but are perfectly capable of surviving in their limited way—even though they do not possess their own machinery for self-replication.

A type of virus, for example, attacks bacteria. These are the phages, the most famous of them being the T_2 phage which attacks the bacterium *E. coli*. The virus injects its DNA into the bacterium, thereby taking over the direction of the bacterium's synthetic work and directing it to make phages instead of its own material. Not all phages live in cells, but they are all synthesized in cells, at the expense of the host's own protein factories.

These then are some of the main units of the cell and their respective functions. The organization of the cell is functional and highly developed, enabling the necessary work to be done with maximum efficiency. It only goes to show what a few billion years of evolution can do.

CHAPTER 2

chemical reactions

The fabric of biochemistry is woven from the ideas of chemistry—ideas concerning atoms, molecules, and their behaviour under varied conditions. The problem is to relate knowledge obtained usually from relatively small entities observed under strictly controlled conditions to the realities of large and complex systems consisting not only of very much larger species, but many more of them, all being influenced by each other. An additional difficulty is that in the living organism the scientist has to accept conditions as he finds them: he cannot employ the usual method of the physical sciences of allowing only one or two variables to affect the results. Despite all these formidable experimental and conceptual difficulties, biochemistry is anchored very firmly indeed in the corpus of chemical knowledge.

Let us take a statement which will be amplified in a later chapter and consider some of its implications: proteins are polypeptides composed of amino acids. We can specify the components of virtually all chemical entities involved in the life processes and often can also describe the way in which they are associated. Before considering amino acids, we must see how they are constituted. Let us assume that amino acids, a class of chemical compounds, consist of atoms, usually of no more than four or five different kinds.

There are over a hundred *elements* and each of these corresponds to one type of *atom*. Thus the total number of combinations among atoms runs into many millions. A large number of these are known and have names, and to make the situation slightly less complicated, large groups of them, classes of chemical compounds, have family names of their own; to save their sanity, chemists have devised a way of systematically naming each and every known compound.

Before considering the components of the large, biochemically interesting entities, we must refresh our memories of small-scale chemicals and look at the atoms composing them. Each type of atom is itself a composite of *sub-atomic particles*, and although the hunt for more and more is the constant and costly business of the atomic physicist, chemists can explain most of their discipline by worrying about three only. Of the three, two are included in the *nucleus* of an atom: the positively charged *proton* and the uncharged *neutron*. Free atoms are electrically neutral overall; therefore to counterbalance the positive charges on the nucleus arising from the protons, an equal number of negatively charged *electrons* are also present in the sphere of influence of an atom.

Since we are not concerned with the details of chemical physics, we can take a few shortcuts and assume that the way an atom will behave chemically is essentially determined by the number of protons in its nucleus and therefore the number of electrons. This number is called the *atomic number*. But there can also be neutrons in the nucleus which, being uncharged, do not appreciably affect the chemical nature of the atom. They do, however, affect its mass. Now it would be very difficult indeed to determine the absolute mass of an individual atom, and in any case the number obtained would be so small as to make it quite impossible to use it for day-to-day calculations. *Atomic masses*, the mass of the number of protons and neutrons in a nucleus, are therefore quoted in terms of an agreed unit. This standard is currently defined, so that carbon containing six protons and six neutrons in its nucleus is carbon-12, which is usually written in the form ^{12}C.

The fact that we can have atoms with the same atomic number but different mass numbers leads to the very important result that atoms which are virtually identical chemically can still possess a measurable difference—that of mass. This concept is described by the word *nuclides*—or in an earlier terminology, *isotopes*. Thus amino acids that make up a protein contain the element nitrogen which can occur with mass numbers of 14 or 15. The atomic mass of an element is simply the average of nuclide masses in a natural sample.

If it is possible to prepare one or more amino acids containing a nitrogen atom of a known mass, for instance, nitrogen-15, ^{15}N, it should be relatively simple to follow its incorporation into a protein and the subsequent fate of this large entity, if we have a technique that will distinguish ^{15}N from all other nitrogen isotopes. Mass differences thus give, at least in theory, a way in which particular atoms can be marked or tagged, the success of marking being dependent mainly on our ability to devise an efficient technique of analysis.

Luckily, a number of isotopes are radioactive: they emit a radiation which can readily be picked up by a Geiger counter or similar instrument. In biochemistry the radioactive isotopes used include carbon, sulphur, and phosphorus.

There are also a number of isotopes that are not radioactive; from the biochemical point of view, 'heavy' hydrogen (deuterium) and ^{15}N are probably the most important. But even here there are methods of separating atoms of the same element having different masses; the mass spectrometer is one instrument capable of

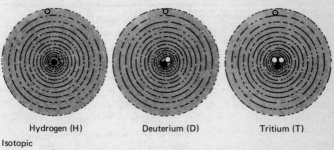

● Protons ⎱
○ Neutrons ⎰ in nucleus
◉ Electrons

| Hydrogen (H) | Deuterium (D) | Tritium (T) |

Isotopic
mass =   1   2   3

Figure 2   Isotopes of Hydrogen

Water in which the hydrogen atoms are replaced by deuterium atoms is called 'heavy' or 'deuterated' water (D_2O). Tritium, unlike H and D, is radio-active and can be used to 'tag' molecules.

recognizing different atomic masses by noting their different trajectories under the influence of electric and magnetic fields.

It is possible, for instance, to feed an animal some material which, when passing through its metabolic processes, acts as a precursor to a protein. If this material contains a tagged atom we should be able to follow not only the buildup of the protein but also its eventual destruction and the re-use of its components. Or we can follow the passage of materials through the cell membranes, from one organ to the next, from ingestion to excretion.

The importance of tagging atoms for biochemical work can hardly be overemphasized. Indeed, detailed studies of metabolism, especially cell metabolism, are to a large extent only possible because these techniques, and the instruments necessary, have become available, mainly since the last war, although tracer studies started in 1920s.

MOLECULAR STRUCTURE

In biochemistry, as in most of chemistry, we are essentially concerned with collections of atoms called *molecules*. There are a number of ground rules of the game of compound-making which take into account that any one atom can form only a prescribed number of linkages with other atoms, and indeed may only have a definite number of close neighbours. The nitrogen of our amino acid, for instance, can only have either three or four neighbouring atoms, and in either case the molecule formed will have specific and known properties.

Possibly the most important rule is that any given molecule is always put together in the same way—it always contains the same atoms in the same proportions. This rule, together with another fundamental law stating that whatever happens to a group of molecules the total number of atoms must remain constant, effectively defines the framework of chemistry and biochemistry.

We noted that atoms in a free state are electrically neutral, and so are molecules. It happens, however, that an atom either in its free state or when forming part of a molecule, can lose an electron or acquire an additional one. The result is that the atom or molecule becomes charged. When a charged molecule is reasonably small we call it an *ion*, as is also the case for a charged atom. If, as is often the case in biochemistry, the entity is very large, it is best referred to as a *charged species*, to distinguish it from its uncharged companions. In considering charged species, it is only

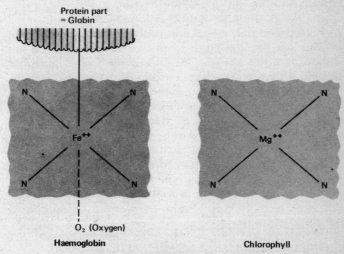

Figure 3 Haemoglobin and Chlorophyll are Examples of Naturally Occurring Metal Complexes

In *haemoglobin,* an atom of iron (Fe) is linked to the four nitrogen (N) atoms of a four-ring structure called porphyrin. This complex is called haem. The iron atom also forms a link with a nitrogen atom of the protein (globin). A sixth bond is formed between the iron atom and a molecule of oxygen or water.

Chlorophyll contains a similar 'nucleus' where a magnesium (Mg) atom is linked to nitrogen atoms in porphyrin-like structures.

necessary to remember that in general the total system must be electrically neutral. If it is not, as in the case of nerve conduction or muscle action, we must have a good explanation to account for it.

There are several variations of these ideas which explain the differences between molecular species. An interesting class of compounds that performs a vital role in enzymes, among other substances, are the *complexes* based upon a metal centre. The metal is surrounded by a number of groups, and the whole may be electrically charged. The importance of a complex is that the assembly—metal and surrounding groups—can act as a unit. But it is also possible to change the groups surrounding the metal without bringing into play the whole complex.

The role of complexes in biochemical systems arises from the possibility that while the total complex, metal and surrounding groups, may be anchored on to an even larger entity, it is still possible, generally speaking, to alter groups in the immediate vicinity of the metal. The best example for this phenomenon is the function of haemoglobin, the oxygen carrier of the blood, which contains the haem group, a complex based on an iron atom. While the haem group is securely fastened to its surrounding protein, the iron atom can take on and shed oxygen.

Similar chemical behaviour occurs with a number of enzymes, where the actual catalytic part relies on the ability of a metal atom, which is part of a complex, to change its surrounding groups while remaining in the same general environment.

THE SIGNIFICANCE OF ATOMIC CONFIGURATION AND CHEMICAL BONDS

Until the advent of the electron microscope nobody had ever seen a chemical compound and it seems at present most unlikely that we shall ever see an individual atom directly. Yet chemistry developed before this time to a great extent because its practitioners possessed excellent imagination: their imagination assured them, as was subsequently proved by experiments, that the formulae they were writing represented only the palest shadow of what was happening in reality. The real nature of chemical compounds can only be appreciated if one thinks of them as three-dimensional entities, having definite shapes, sizes, and even orientations in space. One can adduce a number of reasons for this fact, according to the level of mathematical sophistication to which one aspires. Basically, however, the atoms in any molecular species can hold together only if they are in certain definite spatial relationships to each other. Without this underlying fact, whole areas of biochemistry or chemistry would immediately become mystical gibberish.

The idea of molecular shape has a number of important consequences. A *reaction*, that is, a redistribution of atoms, can take place if the reactants can get within striking distance of each other. If they cannot do this, the reaction will not occur, even if theory would predict it. Such ideas, under the general heading of stereochemical considerations, play an important part when we consider the biochemical significance of a wide range of compounds.

The most obvious examples of the fundamental significance of *stereochemistry* are the elucidation of the structures of the nucleic acids, DNA, RNA, and the proteins. Their properties can only be satisfactorily explained on the basis of certain dimensional arrays of atoms, and indeed, properties could be regarded as a dynamic consequence of static architecture. It is hardly necessary to invoke the complexities of the double helix. We can, for example, consider the action of enzymes in catalysing the reactions vital to the organism. Even when the details of these reactions are not known, it is clear that the ground plan relies on a close, temporary association between molecules formed and broken down according to a detailed pattern. Such an association, a type of complex, is only possible through a measure of stereochemical sympathy between the molecules taking part. This can be made even more specific in the lock-and-key theory of enzyme action discussed later, a theory that can even be applied to the action of some drugs in blocking the effects of bacteria.

The importance of three-dimensional structure can be shown for virtually all molecular action, property, and therefore biochemical function. It also leads to some very interesting problems. A number of molecules—which can be defined by a number of mathematical rules—can exist in two forms that are mirror images of each other: these are called *isomers*. Now it is obvious that mirror images are identical, except that they cannot be superimposed. One should expect, therefore, that chemically they should be equivalent, and in the course of common or garden

(a)

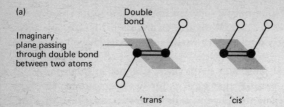

Imaginary plane passing through double bond between two atoms

Double bond

'trans'   'cis'

These are 'geometric' isomers.

(b)

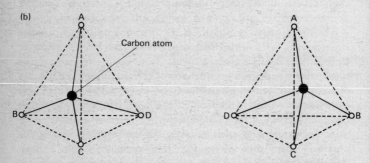

A

Carbon atom

B   D

C

A

D   B

C

3–D Representation of a carbon atom (situated somewhere in in the middle of a tetrahedron) to which four different atoms A, B, C, and D are attached.

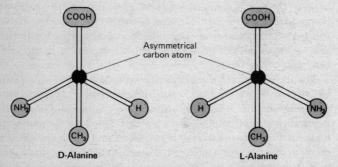

COOH

NH₂   H

CH₃

D-Alanine

Asymmetrical carbon atom

COOH

H   NH₂

CH₃

L-Alanine

An example of the isomers of the amino acid called alanine

Figure 4 Isomers
(a) Isomers are molecules identical in composition but in which the arrangement of the atoms differs.
(b) Compounds of identical composition may be asymmetrical if their arrangements in space are different.

chemical reactions they usually are. Not so in biochemical systems. It seems that one of the mirror-image pairs is usually wholly favoured over the other one in the construction of some biochemically important molecules. For instance, the amino acids making up a protein molecule can exist in two configurations, but it turns out that all proteins are assemblies of only one form. Why this should be we do not know. It does not seem to confer any obvious advantage on the resultant product, although some recent research seems to indicate that it may be more resistant to radiation. One can only speculate that at some time during biochemical evolution a particular protein showed high powers of function. This protein, by a completely random process, might have been constructed of amino acids of a particular configuration. We can further suppose that once this jigsaw dropped in place, its very existence created an environmental effect that tended to favour the emergence of similarly constructed units, so that the current construction of proteins is, in a sense, determined by a particular historical accident.

Although one of the most important considerations, shape is only one factor determining the final characteristics of a compound. The other is the strength of the *bonds* between atoms. These strengths can be measured and the measurements interpreted. All that such data say is essentially that some combinations of atoms are more difficult to tear apart than others, and if we place the compound in a situation where some bonds are likely to break, the weakest ones will go first. This idea does have some rather important consequences. It has been found experimentally, and good reasons were later adduced for it, that certain combinations of atoms are held together very strongly indeed. Such combinations may be molecules in their own right, for instance, methane, CH_4 or diamond, C_n, but they can also be but a small part of a larger molecule. In such a case we call them *groups*: for example, the methyl group, CH_3-. The importance of these closely knit groups is that in some cases we can treat them in an analogous fashion to tagged atoms. In other words, a molecule containing a particular group may take part in a number of chemical reactions. All sorts of rearrangements may take place. Yet all through such evolutions and revolutions, a group stays together and reacts as an entity; its presence can therefore serve as an indication of what happened to the rest of the molecule.

THE CONSTRUCTION OF MACROMOLECULES

The building blocks of biochemistry are firmly based on the sequence:

$$atoms \rightarrow compounds \rightarrow very\ large\ compounds$$

the latter are usually called *macromolecules*. Compounds occupy a central role, because not only are they important in their own right but they also serve as building blocks for the larger macromolecules. Proteins and nucleic acids are among their representatives. Types of macromolecules called *polymers* are well known outside biochemistry: plastics, glass, and clays are some of their important members.

Such a statement is not specific to biochemistry: it belongs to the realm of chemistry. What makes biochemistry so fascinating is that the enormous variety of species that contribute to the

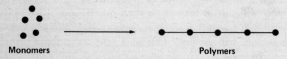

Monomers   Polymers

Polymers are often chains of similar repeating small molecules.

Polymer chains can be intertwined to form coils or helices.

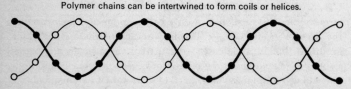

Figure 5 Monomers and Polymers

functioning of even the smallest organism are built up from a relatively few elements and a larger, but still understandable number of classes of compounds. We can, therefore, with advantage look at some of the simpler sorts of compounds and also consider some of the general properties of the giant macromolecules they may produce.

Biochemically by far the most important material is water, H_2O, the formula which even those readers without the slightest trace of chemical knowledge will remember from their school days. Water has a number of interesting properties. From the biochemical standpoint, the most important is probably its characteristic of being a very good solvent; it can therefore be used as a carrier for other material.

A very large number of compounds, irrespective of their architectural details, fall in the category of acids and bases. There are again a number of definitions of these two classes; a reasonably simple one states that an *acid* is a compound that sheds a hydrogen ion (or a hydrogen minus its electron called a proton) with reasonable ease. Conversely a *base* is a molecule that can pick up a hydrogen ion or proton shed by an acid. Acids and bases are important in biochemical systems because in some of the most vital functions, such as cell respiration, a whole chain of chemical reactions gets under way by passing a proton along the chain. It has also been found that certain systems of reactions, for example those involved in the digestion of food, only function when their environment is sufficiently acidic or basic. If the mechanism goes wrong, we may experience the discomfort of indigestion. The idea of acidity and basicity can be made quantitative, and chemists use what is called the *pH scale* to do this. On this scale, which runs from 0 to 14, the lower the number on the scale the more acid the environment, and the more basic the environment, the higher the number. When, say, a solution is neither basic nor acid, in other words *neutral*, it can be defined as having pH 7.

Another way we can look at a number of compounds is to say that they are *oxidizing* or *reducing agents*. An oxidizing agent, at least in principle, is a molecule that is liable to shed an oxygen atom; a reducing agent can take one up. But through arguments by analogy, this idea has been taken several stages further, so that it is also possible to call a molecule shedding a hydrogen ion (an acid) a reducing agent, and a base an oxidizing agent. It is also possible to rephrase the concepts of acidity/basicity and oxidation/reduction in terms of electrons: a molecule losing an electron, equivalent to gaining a proton, is oxidized; one gaining an electron is said to be reduced. These ideas superficially belong to the details of classical chemistry, but their importance to biochemical systems comes through the realization that the energy required by an organism, apart from plants, to grow and indeed to survive, comes from combustion in a way somewhat analogous to our burning fuel to keep our homes warm. Burning is equivalent to oxidation, and therefore a large part of biochemistry deals with the various ways molecules may be oxidized in a living environment.

We started this chapter by saying that proteins are composed of amino acids. There are a number of other smallish molecules that turn up in all manner of biochemical circles: sugars, phosphates, and various types of bases and acids. It is perhaps sufficient at this point only to take cognizance of their existence and wait for more details when we consider their roles in particular systems. It is, however, interesting to note that virtually all of them are eminently suitable for linking up either with each other or some other type of molecule, thus forming large collections or *chains*. If macromolecules are made up of a small number of units which repeat themselves, they are defined as polymers. Macromolecules are, of course, exceedingly large. In order to obtain some quantitative idea, we can consider that the usual run of molecules that are of interest to chemists have a molecular weight of a few tens or a few hundreds. Macromolecules, like proteins or nucleic acids, are characterized by molecular masses of thousands or in some cases millions.

Now although we have firmly eschewed the vitalist theory which asserts that life chemicals possess characteristics that cannot be reproduced in a test-tube, we do believe that when a molecular species becomes very large indeed it does have some properties that are essentially dependent on size and complexity. There are several good reasons for this.

We have noted that all molecules have a three-dimensional shape. When they join up, they still hold on to their orientation so that long chains are never smooth and frequently twist about in a most theatrical manner. For example, the shape of the amino acids making up a protein demands that protein molecules should be helical, and in the same manner the shapes of its components determine the famous helix of the nucleic acid DNA. It is im-

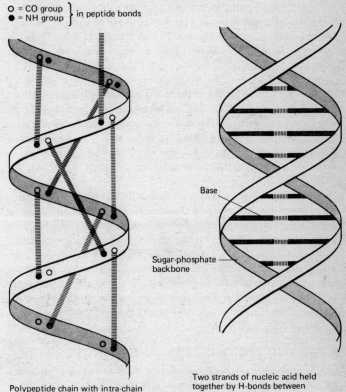

O = CO group ⎫
● = NH group ⎬ in peptide bonds
              ⎭

Base

Sugar-phosphate backbone

Polypeptide chain with intra-chain H-bonding

Two strands of nucleic acid held together by H-bonds between complementary bases

Figure 6 Hydrogen Bonds
Hydrogen bonds help to maintain large polymeric structures, e.g. they maintain the coiled structures in proteins and nucleic acids.

portant to appreciate that these shapes are purely the results of the spatial orientations of their components. Because giant molecules are not strung out in a line, sections of them are so near to each other that one part of the chain can influence the behaviour of another. It is possible to sort out the nature of most of these influences: one very important feature is described as *hydrogen bonding* and consists of a weak influence exerted by a hydrogen atom already forming part of a molecule. Although the effect of any one hydrogen bond is very small, there are usually a great many of them, so that the total force is appreciable. Thus, for instance, the stability of both proteins and nucleic acids is greatly enhanced by hydrogen bonding across the twists of the spiral.

Because polymers are so huge, they can have a variety of different groups attached to them. As we have seen, some groups at least have the ability to act in a semi-autonomous manner even when they are part of a larger assembly. One can imagine, therefore, that when a variety of groups are attached to the backbone of a chain, we can have all manner of effects since, in addition to acting on their own, they must also influence each other along the chain.

Arguments of this sort can be elaborated to high complexities, and indeed the existence of giant molecules is a good source of both experimental and theoretical headaches to those working with biochemical systems. Sometimes they are difficult to handle and suffer damage and alteration and their very size makes them a nuisance when it comes to measurements. Yet they are of the utmost importance since their size and ability to carry out a number of different functions make them pre-eminently suitable for the biological systems in which they play a part.

DYNAMIC CHEMISTRY: THE REACTION

Up to now we have been considering molecules rather as architectural curiosities, but the most interesting part of chemistry deals with what happens to molecules and to ions in action. This generally goes under the name of chemical reactions, although there are a number of additional possibilities—osmosis, for instance.

The number one consideration is that irrespective of how many molecules or ions take part in a chemical reaction, and however complete is the resultant rearrangement of the atoms taking part, none of them disappear or appear from nowhere: all have to be accounted for. If this is apparently impossible, we have gone wrong in interpreting our results and have to start again. Rule number two refers to the number of different atoms any given atom can associate with. The associative power of an atom used to be described through the idea of *valency*; thus a carbon atom had a usual valency of four, meaning that it could form a compound with, say, four atoms of hydrogen, each of unit valency, with two atoms of oxygen, which has a valency of two, and so forth.

In classical chemistry, or indeed in most chemistry outside biochemical systems, it is usually sufficient to understand the course and products of one or a small number of related chemical reactions. In biochemistry, matters are usually more complicated. This arises from the fact that the product of one reaction can often become the raw material for the next reaction, and so on. Thus one can obtain a long chain of reactions, each of them being dependent on the previous one: the corollary is that if any given step goes wrong, the whole chain stops functioning correctly.

We can further consider such closely tied systems in terms of feedback theory. Let us take a relatively simple scheme, consisting of two steps only—a conventional chemical reaction. If starting material A turns into B, the production of B will be equivalent to the rate of disappearance of A. We may, of course, have a situation where, to all intents and purposes, B has very little or negligible effect on the extent and speed of its own production: for instance, if it is taken away as fast as it is formed. Two other situations may also occur. The presence of B may slow down the disappearance of A by some mechanism we need not specify; this will amount to *negative feedback*. In other words, the more B that is produced, the smaller will be the additional amounts made, and the system will gradually come to *equilibrium*. A number of such systems have an important effect on, say, the rate of hormone production during the menstrual cycle.

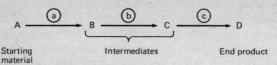

Starting
material
Intermediates
End product

Each step is catalysed by a specific enzyme, a, b, and c. Metabolic pathways often work in both directions and the specific enzyme catalyses the reaction in both directions. Reversible systems are symbolized thus:

**Figure 7   A Metabolic Pathway for Conversion
of Substance A to Substance D**

On the other hand, the presence of B may stimulate its own production. This would be positive feedback and the mechanism would undoubtedly be catalytical. This is called *autocatalysis*, since B is produced by the reaction being catalysed. A *catalyst* is usually described as a molecule or some other agency that speeds up a chemical reaction without itself suffering any change by the end of it. It is important to bear this qualification in mind: a catalyst does indeed take a very active part in the reaction it speeds up, and usually undergoes very definite changes during the reaction. These changes are reversible, so that, theoretically at least, the unchanged catalyst can be recovered at the end. Because a catalyst

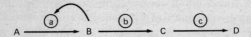

When B reaches a certain concentration (amount), it can slow down enzyme a and thus limit further conversion of A to B. This is called 'product inhibition'. Alternatively, D inhibits the first step of the sequence. This is called 'end-product' inhibition.

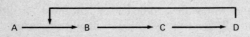

**Figure 8   Metabolic Pathways and Negative
Feedback**

does take part in the actual process of reaction, it can be made ineffective if we interfere in some way with its ability to take part in the reaction. This is normally described as poisoning the catalyst. Large areas of biochemistry have been explored by poisoning specific catalysts, mostly enzymes, and observing what happens when a particular reaction is blocked by these means.

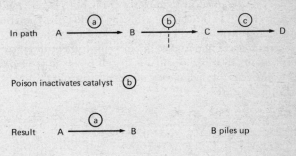

Figure 9 Metabolic Poisons

It has also become clear recently that some very important reactions, such as oxidative phosphorylation in cell respiration, take place on the membranes of the cell; these have a great deal in common with the industrial reactions occurring under the influence of surface catalysts, for example in the production of petrochemicals. As we have suggested, a reaction cannot take place if the reactants cannot get within striking distance of each other. Catalysts, such as membranes that play a part in surface reactions, and some types of enzymes, have essentially the function of enabling the reactant molecules to get very near each other. This they do by having shapes that fit well those of the molecules involved. In the living cell these surfaces may also be adjustable, so that their shapes are modified according to the reactions taking place on them.

Another membrane effect of importance is that of *osmosis*. If a solution of a material and the pure solvent are accessible to each other, there will be a tendency for the dissolved molecules to spread themselves in both the solution and the pure solvent, or as we say, for solute concentrations to be equalized. This tendency can actually be measured if one separates a solution and a solvent by a membrane permeable to the solvent but not to the dissolved molecules. The tendency of the dissolved molecules to pass into the pure solvent is equivalent to the force of the pure solvent tending to force its way into the solution; this can be described in terms of pressure, and measured. The pressure is called *osmotic pressure*. It has several roles of importance in biochemistry. For instance, the concentration of various essential materials is kept at the required levels in the cells through the use of membranes and the operation of osmotic pressure.

SPEED, ENERGY, AND ORDER IN CHEMICAL REACTIONS

Let us return to the question of chemical reactions and consider some implied consequences. All reactions have certain fundamental characteristics in common, although if the reaction system is complex enough a very great deal of work is needed to sort out the individual components.

Although most reactions involve at least two reactants to give one or more products, we can consider an even simpler case where one compound, say molecule A, by some mechanism becomes two products, say B and C. There are a great many examples for such a case: the dissociation of a gas such as hydrogen iodide, HI, being an obvious example.

$$HI \rightleftharpoons \tfrac{1}{2}H_2 + \tfrac{1}{2}I_2.$$

The first point to notice is that the reaction itself proceeds at a given rate, that is, it has a *velocity*. This can be accurately measured by a variety of methods, usually by the disappearance of a reactant or the appearance of a product. Thus, if a reaction proceeds long enough, we should expect all the reactants to disappear—but we should be wrong. In actual fact, very few reactions go to completion: after a certain time the number of reactants forming products equals the number of products which revert to reactants. When such a condition is established, we say that the reaction is at *equilibrium*, and can express it as a ratio of products to reactants. Thus the existence of equilibrium implies that both forward and backward reactions are going on at the same rate, and by the same token we can say that the measured velocity of a reaction is in reality the summation of forward and backward reaction velocities.

Now it is possible to shift the position of equilibrium in a number of ways, for instance by altering the temperature or the pressure of the system. Such a procedure would have little chance of success in a living organism; pressure and temperature changes are severely limited by the overall requirements of the organism. Thus if we raise the temperature more than about four degrees Celsius (the term Celsius is now used in the scientific world in place of Centigrade), a warm-blooded animal will die, so the exercise would lose most of its purpose. However, even if it is impossible to alter the position of equilibrium, it is possible to change the absolute amount of products or reactants. Bearing in mind that equilibrium implies a ratio between products and reactants, if the total amount of one of these is altered, the amount of the other automatically follows suit.

The consequence of this mechanism is that by piling on material on either side of a reaction equilibrium, the concentration of material can be shifted to the other side. This process is used a great deal by biochemical systems, since it can be employed as a relay if one reaction is used to inhibit or drive forward another.

It is important to realize that a catalyst has no bearing whatsoever on the position of equilibrium, only on the speed with which the system reaches it. Thus by supplying a catalyst to a reaction system we shall not be getting a greater amount of products, but what we do get will be available more quickly.

Indeed, the confusion between speed and point of equilibrium has long caused unnecessary difficulties not only to students but also to some practitioners of the subject. Yet there are even two separate approaches in chemistry to reactions: kinetics and thermodynamics. In *kinetics* we are only concerned with the speed of reactions: our comparisons of speed are expressed in terms of the so-called kinetic constants. *Thermodynamics*, on the other hand, deals with equilibria, order, and energetics, and the question of speed and indeed of time does not enter into it. Thermodynamics tells us if a reaction can conceivably take place and if it does, where the position of equilibrium will be. It does not say anything about how long it would take to reach it; if it takes a million years, that is just too bad for the experimenter.

Apart from predicting the position of equilibrium, thermodynamics gives us two further important clues: energy and order. All chemical reactions deal with transfers of energy between individual molecules. The very reason that reactions take place at all is that the new arrangement of atoms is in some way energetically more satisfactory than the previous one.

In the same way, therefore, that we can regard potential energy as driving the fall of an object from a height, we can also assume the existence of chemical energy to drive a reaction. In just the same way that a moving object under the influence of its surroundings produces work, so a system of chemical reactions may also be a source of work or energy—or may require the input of energy from its surroundings. The amount of energy a reaction system supplies or uses is usually measured by an arbitrary set of thermodynamic functions under the name of *Gibbs free energy*—that can incidentally be correlated with a variety of other properties of the molecules involved.

The important point to note is that when we talk about the energy of a reaction—and biochemists talk very frequently about it—we must bear in mind not only the compounds taking part in the reactions involved, but also their environment. For example, when certain molecules shed a group to their surroundings, such as a phosphate group, energy is liberated. But this liberation of energy is just as much dependent on the molecular environment as on the molecule itself. A further obvious deduction must be that reactions requiring energy for their start or continuation must have the energy supplied from somewhere; and analogously, reactions liberating energy must have some method of getting rid of the accumulated energy.

Lastly, we should also consider the question of order. In chemistry this is expressed by talking about the *entropy* of the system. Basically, the concept of entropy is a mathematical formulation of the fact that left to themselves most systems tend to get more disorderly. In order to maintain or increase order, energy must be supplied to the system. This energy must eventually be derived either from energy made available from chemical reaction or by some other source, usually heat.

CHAPTER 3

functional specialization of cells and tissue formation

The idealized cell we considered is a delightful piece of scientific generalization. It allows the description of major components and functions and the introduction of reasonable arguments suggesting that accurate and delicate mechanisms must be responsible for the position and function of its components. The only difficulty is that the idealized cell, like idealized social systems, does not exist. When we begin to consider real systems, the perfection of the imagination must give way to the necessities of reality, and immediately we are faced with all manner of exceptions and special cases. Yet it is the burden of science to describe the real world and to extract from the particular the laws underlying the general: the task of biochemistry is to answer the innumerable questions posed by the living organism. The necessity of facing the challenge of living matter and constructing the intellectual jigsaw puzzle from its disparate experimental pieces has made all scientific endeavour, and latterly especially the biological sciences, such an absorbing field in which to work.

The point of departure of our inquiry must be the recognition that in all living organisms many groups of cells, or organs, are palpably different. The recognition of these differences is lost in the depth of time, but it must have been clear quite early in the emergence of animal life that certain portions of plants or edible animals taste better than others, and by the time we arrive at the dawn of history a reasonably detailed knowledge of anatomy is a matter of record. From then on, knowledge progresses as a function of available, or sought after, experimental methods. The knowledge of anatomy becomes better and more detailed, although our physiological understanding, that is, what the various pieces of the human and animal apparatus do, lagged far behind and indeed is hardly complete even today.

Nevertheless, the idea that human or animal function could be explained on the basis of the physiological actions of its constituent parts was well current by the eighteenth century, although those explanations more often than not appear to us distinctly bizarre. Today we possess a far more accurate understanding of how groups of cells, the tissues, carry out their tasks, although undoubtedly many of our explanations will be regarded as wrong in the future.

As our knowledge of individual cells has increased, we also have to contend with another problem: how to correlate what an individual cell does with the function of the tissue. Although in most cases the overall flow of chemicals into and out of an organ is known in considerable detail, we have to marry this knowledge to the precise nature and sequence of the chemical reaction occurring in an individual cell forming part of the whole system; our correlation between large-scale events and molecular-scale reactions is still imperfect.

Our second, and even more complex, problem concerns the question: How do specialized cells become specialized? The solution of this problem is going to be one of the most significant areas of growth in the biochemistry of the future, and its solution will equal in its consequences the decoding of the genetic and control mechanisms of the cell.

What precisely do we mean by specialization? The underlying idea must be that the properties of a tissue are the expression of cell function, and therefore a muscle cell, for example, will show contractile properties as behoves a component of a contractile organ. In the human body there are about ten million million cells, all of roughly the same constitution, all of them carrying out certain reactions. But in addition certain classes perform special functions. For instance, red blood cells carry oxygen, yet they do not use it themselves. Muscle cells, on the other hand, do require large quantities of oxygen for the oxidation of the sugars that gives them the energy necessary for contraction and thus for the performance of work. Other cells have other specialties: liver cells make albumen, plasma cells make antibodies, and we could extend this list to many thousands of examples. In all cases one could describe the underlying principle in terms of the old Marxist idea of 'from each according to his ability, to each according to his needs.'

It would not be misleading to think about cell specialization, at least as a first approximation, in terms of a sociological model. In a reasonably advanced society most people perform specialized functions: managers manage, salesmen sell, lawyers argue. Underlying these specialized functions are a wide range of general functions common to all members of a society. For instance, they are all consumers and many of them drive cars. In the same way, all cells partake in certain functions common to them all, and in addition are able to carry out tasks which are particular only to them.

There is also another parallel between social and cell organization. In the same way that nobody today can be a self-contained unit, able to work and exist without reliance on those around him, and being relied upon by others, so it is often best to consider groups of cells as being integrated into the system as a whole rather than being only responsible for certain well-defined operations. In particular cases it is therefore profitable to consider assemblages of cells, the tissues, or groups of tissues working together, the organs, or systems of organs, the individual. Such a scheme of increasing size does correspond, in a reverse order, to the historical development of the sciences dealing with living things and our possession of techniques suitable for probing smaller and smaller effective units. The study of individual cells offers many advantages; they can, for instance, be isolated and their reactions rigorously followed. All the time, however, it is necessary to bear in mind that the cell in its assembly will probably have special constraints imposed on it because of its position in the system.

Cell specialization, although most obviously high up on the evolutionary tree, does take place in even the most simple

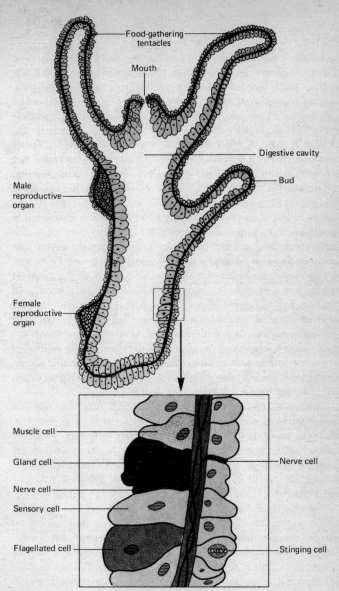

Figure 10   The Hydra
The specialization of cells

organisms. Considerable cell specialization, and therefore division of labour, occurs, for example, in the hydra, where one can distinguish male and female sex tissues, digestive tissues, nerve cells, stinging cells, glands, and muscles.

At this point our analogy begins to break down. It is impossible to consider a social framework without taking into account competition between single individuals or groups of individuals; but in considering cell systems, we have to focus on co-operation between units rather than on competition. In terms of economics, co-operation is the most viable arrangement in a stable system. If groups of cells can specialize in certain areas, for example in processing food for the whole organism, other groups can provide some other equally necessary service. Such an arrangement avoids duplication, and presumably over an evolutionary period the specialized groups become better at their own specialty. There are, however, certain inherent dangers. If the total organism is damaged, the extent of damage can be much larger if some vital function requiring specialized cells is damaged, rather than if a large number of unspecialized cells are killed off, presumably leaving enough of their colleagues to take over. As we go up the evolutionary scale there is a definite decrease in the ability of organisms to regenerate, and it could be argued that this is connected with their ability to avoid situations where regeneration would be necessary.

Regenerative powers also vary a great deal between organs. The most important example of an organ which cannot regenerate itself is the brain. It is estimated that the human brain loses about 10 000 cells every day, but it does not have the ability to regenerate itself—a physiological fact which provides rather melancholy justification for the dictum about dying each day by small degrees.

The effects of this prodigious rate of dying are staved off for three reasons. Our current understanding of brain function argues that all the functional units of the brain are heavily duplicated, rather in the manner of duplicated braking systems in cars or electronic circuits in rocketry. We therefore start off with a number of functional brain units in excess of our needs and can afford to lose some along the way. We also know that areas of the brain are responsible for particular functions: there are, for instance, the visual, auditory, and locomotor areas, all of them enjoying a degree of autonomy. Damage done to any one of these areas is not generally lethal. Finally, one area may also take over the tasks of another: for example, polio victims may learn to utilize a different set of muscles, so that brain commands switch over to a different set of outlets in order to carry out a set of tasks.

On the other hand, we can take red blood cells as examples of particular cells being made continuously, in this case at the rate of about 10 000 a second. The half-life of the blood cell population—the time taken for half the members of a group to die off—is about fifty to seventy days, and no individual cell is therefore likely to spend more than a few months as a viable unit.

There are a great number of examples of particular types of cells able to regenerate themselves. If even a relatively large piece of liver is cut off, the tissue will grow again to its original size, no larger or smaller. Not only must we assume the presence of factors making regeneration possible, but also additional ones regulating the amount of regeneration taking place.

The cells of the outer layer of skin are also members of a class of cells that are dying all the time. Dead cells are replaced by new ones surfacing from deeper layers, where they are actually created through cell division. Once they come to the surface they no longer divide; but on the other hand they start making keratin, a fibrous protein ('horny' cells).

Regenerative powers are greatly enhanced among lower vertebrates and non-vertebrates. The salamander is the usual example for a vertebrate that can lose a limb and not only live to tell the tale but also grow another one in its stead—a peculiarity that no doubt had some effect on the folk myths about its magical powers. Lower down the scale, regeneration becomes even more common and also probably important from the point of view of preserving the species. Flatworms, for one, can lose large chunks of their bodies without coming to very great harm, a situation which must be reassuring from the flatworm's point of view.

The overall situation we therefore have to account for is that there are large agglomerations of cells, the tissues, which carry out specialized tasks. There has to be some pattern according to which specialized cells are constructed and which also makes sure that cells of a particular type all end up at their appropriate places. There has to be both cell differentiation and morphogenetic movement, that is, the migration of groups of cells to their ultimate destination.

The necessity to meet these two criteria immediately introduces difficulties when we try to describe a control mechanism powerful and subtle enough to accomplish these tasks. Every organism starts with a single fertilized cell. Such a cell divides, the two daughter cells divide again, and so on. Very early during this process the new cells formed are no longer generalized units: they have become differentiated and form the basis of future organs. Not only has differentiation taken place, but these new cells are in the right position; control has been exercised over the shape of the growing organism.

In the human embryo, for example, after about a few hundred cells have been formed, one can already perceive the future of the different parts. After a few cell divisions of the fertilized egg, a hollow ball called a *blastula* is formed from a single layer of cells. The functional advantage of such an arrangement is that nutrients can easily diffuse into and waste products out of the blastula. As the embryo grows, the ball begins to fold on the inside, and invagination takes place—a funnel-shaped structure appears at one end and eventually reaches the other pole. This process is called gastrulation and this stage of embryonic development is called *gastrula*. The cells arrange themselves in concentric rings around the interior of the ball. The funnel-shaped invagination is the beginning of the gut system. From this stage on, we can follow the eventual destination of the cells in the embryo. The outside layer differentiates into three layers: the *ectoderm*, the *mesoderm*, and the *endoderm*. The ectoderm becomes the epidermal ectoderm, giving rise to the skin, teeth, hair, and nails, and the neural ectoderm, the forerunner of the central nervous system: the spinal cord and the brain. The mesoderm becomes the muscle tissue, the excretory and reproductive organs, the skeleton, much of the digestive system, and the circulatory system including the heart. The endoderm layer becomes the internal lining of a range of organs, including the digestive system and the bladder.

By the time the gastrula stage is reached, we have therefore arrived at a stage where all cells are partly differentiated. Presumably they are also already in possession of whatever factors are necessary for further differentiation, and in a sense the epidermal ectoderm is already the skin that will indubitably form with the passage of a certain amount of time. This is an argument along one axis of time, the future. Passage into the reverse direction of time would logically have to lead us to a state in which each cell was truly multipotential, capable of performing every task, or capable of developing in any conceivable direction. Unfortunately, apart from its self-contained logic, this argument has no basis in physical fact. Nobody as yet has been able to show such a truly multipotential cell, so that the question of when, if ever, a cell is multipotential, and if it ever is, when it ceases to be so, is at present unresolved.

The question we therefore have to answer is that having started with one fertilized cell, a certain number of divisions later we have to account for a large number of specialized cells, which in the nature of development become even more specialized and also congregate in their appropriate places. Obviously, the instructions that define these happenings must somehow be stored in the genetic blueprint the fertilized cell receives, but to date we have to admit our ignorance of how it is coded or put into practice. We shall see in later chapters that a great deal is known about how the cell receives its genetic information, how this information is put into practice in the functional life of the cell, and indeed, how the transmission of information is controlled to enable the cell to be specialized. In all this we are describing a situation as it is, and not how it has come about. It is the second of these problems that will, we suggest, give some of the greatest headaches to biochemists in the future, and also where some of the greatest discoveries will be made.

At this time we can only make some reasonable guesses about the nature of the information that must be imparted to cells in order to make them develop in the appropriate manner in the necessary place. How is it necessary to communicate to a cell the command to obtain certain properties and then to congregate to form, say, a finger? And taking this argument one step further, how can we describe a series of logical links between a relatively small collection of atoms in the fertilized egg and the stupendous complexity of the final organism?

One way would be to describe all the necessary properties of such an end product and derive some sort of sequence which could end with a code directing all steps of development from the beginning. This, after all, is the sort of information that an engineering drawing conveys, and one could argue that, although a living organism is far more complicated than a motor car, the difficulties are a matter of quantity rather than of degree.

Despite the superficial attractions of this idea, our current thinking is that it simply would not work. We have no means of connecting up the necessary steps starting from the single cell, through all stages of differentiation and development to the final organism. Something simpler is required.

A simplification may be introduced by considering the nature of our instructions to build an organism. Instead of describing the finished product and therefore the patterns necessary for its construction, let us try to describe a sequence of relatively small actions.

One could specify the formation of the blastula in terms of detailed instructions to every cell about what it will do at any point in time. However, there is an even simpler way to do it. Let us first of all instruct some cells to be stickier than others. When sticky cells meet, they will tend to congregate around each other. Let us also make each cell have a number of very fine tendrils, called *pseudopoda*. When these tendrils touch a surface, they tend to drag the cell toward it. If they do not touch, nothing happens.

If we stipulate these two requirements, we can show that we shall indeed obtain the type of arrangement shown by cells on the inside of a blastula, even without any further control. In other words, instead of specifying the pattern itself, we have specified certain properties and left the rest to chance and geometric expediency.

One can demonstrate how this sort of idea works in practice on a jigsaw puzzle. If the pieces are so constructed that whenever specific units meet they hold on to each other, shaking the pieces together will eventually compose the entire puzzle, although this process may take a considerable time. Small actual puzzles of this type have indeed been constructed and there seems no logical reason why, at least in theory, their size could not be considerably increased.

There is a further argument against very large numbers of precise specifications. If every action by every cell would have to be specified, it is almost certain that the number of errors would noticeably increase. As demonstrated quite dramatically by highly complex systems in space research, or in computers, when we increase the number of units that have to function correctly before the total system becomes viable, the chances of just one malfunction increase correspondingly. In a system containing ten million million units, the chances of individual errors must be very high, despite duplication and the organism's equivalent of fail-safe procedures. Consequently the chances of developing a viable, well-functioning organism must be low. But in fact there are relatively few errors committed in nature. We must argue therefore that there can only be overall guideline specifications and that the rest is left to chance.

Having opted for simplicity in planning, we still have a major difficulty to explain. By all reckoning, even the simplified plans necessary to build a complete living organism must be complicated in the extreme. Where do they originate? In what manner are they transmitted through the generations? Is there some mechanism which allows for changes in form and function, so that the proven processes of evolution may take place? Again, we must point out that although a number of theories do exist, there is as yet no secure knowledge on these questions.

It is now generally acknowledged that the only, or possibly the main transmitter of inherited characteristics resides in a very large molecule, called DNA, the functions of which will be described in a later chapter. In essence, the DNA molecule has the power to regulate the protein synthesis of a cell and therefore its characteristics. It is also known that when a new cell is born through the division of the parent cell, the inheritors both receive a full complement of genetic information. (This statement is true for all but germ cells, where only half of the genetic information is carried.) Now in all organisms of a given species all DNA material is the same—there does not appear to be any obvious way in which DNA can control whether a given cell in an embryo will become part of the skin or the kidney. There is also the observed fact that cell differentiation and migration start to take place before these cells have started to make protein, and therefore before the DNA could have used its regulatory power. Thus there seems good reason to assume that either the DNA molecule is not alone in controlling differentiation and morphogenesis, or if it is, control cannot be exercised in a simple manner.

Structural differentiation does seem to require chemical help from other tissues. It was shown, for example, that the formation of the central nervous system from the ectoderm requires the presence of the mesoderm, by culturing pieces of neural ectoderm tissues with pieces of mesoderm in a solution in a suitable nutrient bath. The results were pieces of brain tissue which, at the time the experiment was performed—in the 1930s, certainly seemed most extraordinary. As we suggested before, since that time the search for differentiating agents has grown apace and will no doubt increase further in the future.

We shall be looking in somewhat greater detail at the biochemical nature of inducers at the molecular level in a later chapter. It should perhaps be sufficient to suggest at this stage that although our ideas on the physical reality of inducers are by no means clear at present, this function seems to involve not only the DNA molecules, the genetic information-carriers, but also several other classes of huge molecules, notably the RNA's and possibly proteins.

Having looked at some of the ideas current about the problems of cell differentiation and tissue formation, we shall now consider some examples of the end product of these processes, in short, what

sort of specific functions can we expect from organs built mainly
from specialized tissue. Our discussion must, of necessity, be exceedingly exclusive, since many rather large books are in existence
dealing with nothing else but the description of the purpose and
functions of organs in living organisms. We hope, however, that
our examples will give some idea of the intricate and exciting way
in which specialization can be expressed in the function of organs.
We shall consider three large and well-known systems: the
digestive system, the kidney, and the liver.

In the same way that architecture can be considered to be an
expression of function—the soaring pinnacles of a cathedral or the
stark efficiency of a concrete office building—so chemical function
on a large, visible scale can be seen as an expression of the identity
of the cells taking part in it. The three examples we shall be considering are all parts of the complex metabolic mechanisms which
are basic to the maintenance of life. We shall be looking later in
some detail at the various metabolic pathways used by the living
organism, especially at the intriguing chemistry that goes on under
their aegis. At this point it is sufficient to argue that the function
of metabolism is to create the necessary ingredients and energy
from the intake of food to maintain and repair the organism and
to serve its growth. In anabolism we are looking at the build-up
of the required ingredients: the mirror-image processes, the
catabolic pathways, are concerned with the breakdown of products, be they molecules too large to be used directly, or noxious
products which have to be broken down and swept out of the
system.

The term 'pathways' is used purposefully: these are complexes
of interlinked reactions, where the product of one step triggers off
the necessary following reaction, and so on. In addition, the
products of any given stage also exercise a feedback function: their
absence or presence in given quantities governs the amount of
material made in the preceding steps. These linked reactions take
place in arrays of specialized cells which congregate to form tissues
and organs. Thus we have the liver, one of the chemical factories
of the body, the digestive tract, which provides the necessary
functions to transform the gross nutritional intake into usable

units, and the kidney, which not only pumps away impurities after they have been brought to a suitable state, but is also concerned with the most vital function of maintaining the correct water balance of the system, on which the function of all other cells is dependent.

THE DIGESTIVE TRACT

The gastro-intestinal tract in man, as in a wide range of animals, is designed to break down large chunks of assimilable substances into portions so small that they can be worked into the system, into cells and circulation, or made to dissolve, generally in water. The tract has to deal with three large groups of dietary chemicals: proteins, fats, and carbohydrates.

Protein digestion The aim of protein digestion is to break down the large protein molecules into their constituent parts, the amino acids. Amino acids are then either reassembled into different proteins required by the body or passed further along the metabolic pathways.

The enzymes which break down proteins, the so-called digestive enzymes, have the distinction of having been one of the first extracellular materials so recognized. This came about in some rather grisly experiments performed by the Abbé Spallanzani in 1783. The good Abbé fed kites pieces of meat enclosed in little wire cages. After some time, the birds, not unreasonably, regurgitated their unappetizing meal and Spallanzani found that the meat was liquefied. Not only that: if the liquefied material was brought into contact with fresh meat, the latter also started to liquefy, although more slowly. The slowness of the process was put down to the temperature being lower than that inside the animal's body—a suggestion made by Réaumur, the French scientist better known for his temperature scale, still occasionally used in Europe.

The agent that liquefied the kites' meat is now known to consist of the enzyme pepsin, a major constituent of gastric juices. Pepsin is the first line of attack on large protein molecules: it is a so-called endo-peptidase, attacking internal linkages in the protein and thus producing smaller peptides and even some amino acids. When the digestion is not working, pepsin exists as a precursor, called pepsinogen, which one can imagine as consisting of pepsin and an inhibitor molecule. Under the influence of acid conditions, pepsinogen is transformed into pepsin and the now inactivated inhibitor. This reaction is reversible and controlled by the acidity of the environment. Under alkaline conditions, all the pepsin exists as its precursor, but once acid conditions are reached, from less than pH 6 down, pepsin is formed and from then on the reaction is catalysed by the very presence of pepsin.

After pepsin has produced smaller polypeptides from proteins, a further range of enzymes takes over in the small loop of the intestine, the duodenum. Trypsin and chymtrypsin, produced by the pancreas, attack the protein fragments at different points of their structure. Yet again, trypsin is secreted by the pancreas as the inactive trypsinogen and is converted to trypsin through the action of an enzyme called enterokinase. Once trypsin is formed, it has the power to catalyse the reaction leading to its own production in a manner analogous to pepsin.

The principle of producing an inactive ingredient that is activated, or triggered, by some other chemical as and when required has a number of implications in biochemical systems, not only in digestion. For instance, recent work tends to show that insulin, a hormone controlling the sugar level of the organism, is also produced in the form of a precursor: in this case a molecule rather larger than insulin. Subsequent reactions chop a sequence out of this molecule, and the two remaining parts become bridged through sulphur atoms to become active insulin.

In the large intestine another set of enzymes, called peptidases (e.g. amino peptidases and carboxypeptidases) finish off the complete breakdown of the peptide molecules into their constituent amino acids. These latter are absorbed into the lymph and are distributed to the various tissues of the body through its circulation. The various sequential steps in these processes are triggered off by the entry of food into the small intestine, where acid secretion signals to the pancreas. A substance called secretin acts both on the pancreas and the liver; the first is activated into secreting the various enzymes required by the digestive juices, the latter into producing bile. In the stomach, the arrival of food starts the production of gastrin which in turn regulates the rate of gastric secretion.

Fat digestion   The digestion of fatty substances, the lipids, is more straightforward than the equivalent process for proteins. The breakdown of fats is carried out in the gut by enzymes called lipases. This process is facilitated by emulsifying agents (the same type of substances that take a prominent part in the making of haircream and mayonnaise) which are secreted by the liver. These agents, the bile salts, also make the degradation products of the lipids more water-soluble, so that they can be absorbed by the blood stream and carried, in the form of fatty droplets, to their storage places in the adipose tissue. To the discomfort of most of us, a great deal of fatty materials are stored this way, and only broken down by hormone-sensitive lipases when the occasion arises for the constituent fatty acids to be used in the system's general metabolism.

Carbohydrate digestion   Our carbohydrate, or sugar intake, consists mainly of starch, glycogen, and the sugars sucrose and lactose. In addition we also take in an amount of cellulose, but having no enzymes to cope with it, cellulose passes through the system and only provides the roughage beloved by breakfast cereal manufacturers.

Starch and glycogen are polysaccharides: they are polymers composed of sugar units. Starch consists mainly of the unbranched polymer of D-glucose, called amylose, and the branched polymer, amylopectin. Glycogen is another polymer, even more branched than amylopectin.

The role of digestion is first to break down the sugar polymers into their constituent units, and then to transform these units, or units ingested on their own, into a form suitable for use by the organism. Sugar absorption may take place by a simple absorption mechanism, or by processes analogous to the ones used in metabolic transport.

THE LIVER

A fair-sized organ situated in the midst of the abdomen and weighing about 7.7 kilogrammes in man is one of the most important chemical factories of the animal organism: important because of its action in the conversion of food and because, like a physician on permanent emergency duty, it deals with the elimination of toxic materials arising out of life processes. It is also an organ we tend to treat with a great deal of disrepect; we overload it with fatty food and alcohol as a result of which a great many of us finish up with livers smothered in layers of fat or, if maltreatment has gone really far, with cirrhosis of the liver—the signal that the liver has had enough and cannot cope any more.

If the maltreatment of our own livers is thoughtless, that of some of our domestic animals is calculated. The pâté de foie gras best loved by all gourmets comes from Alsace, and the method used to give a goose a really large and tasty liver is literally to stuff it full of food and not to allow it to lead an active and energetic life.

Whether unconsciously or not, the liver and its functions certainly loomed large in folk myth before the somewhat more complex reality became evident. The liver, indeed, was regarded as the seat of the soul, and bile, one of its most important secretions, a notable factor in a man's psychological make-up. All this is preserved in our language: the Greek word for bile was chole; choleric and melancholic still testify to the importance of these magic juices, as does lily-livered, beloved by Shakespeare, who together with his contemporaries assumed that thin blood, whatever that may be, comes as a result of kidney dysfunction.

One can summarize in brief the functions of the liver by saying that it acts as a storehouse of energy for metabolic processes, produces major blood components (serum albumen and fibrinogen), and acts as a scavenger to render harmless the potentially lethal chemicals thrown out by the various metabolic processes, or ingested for good or bad reason.

The liver's function of storing energy is connected with the warehousing of sugars, deposited in the form of glycogen, as in muscle tissues. One can look upon sugar as a form of energy carrier: its oxidation, helped along by appropriate enzymes, releases energy which then can be used for other chemical reactions. This process is used, for example, in the contraction of muscles.

Glycogen can be broken down to glucose, which in turn is decomposed into carbon dioxide and water. Glucose is either used in the liver itself or sent along to some other part of the body where further energy conversion takes place. It is important to notice that although we talk about glucose as an energy carrier, it has no special attributes in itself: rather it is capable of taking part in certain reactions the end results of which increase the energy available to the system.

The final breakdown of fatty acids also lies in the liver's domain. The digestive system, as we have seen, breaks down the lipids contained in the food intake into fatty acids. The fatty acids are carried by the circulation into the liver, where they are further broken down by a series of chemical reactions which are collectively referred to as the beta oxidation cycle. The end results are units containing two carbon atoms each, which can either be broken down even further or used as building blocks in subsequent processes. In common with virtually all such reactions, the liver's chemical activities are catalysed by enzymes.

All these processes are fundamental to the body's maintenance, and under normal circumstances the liver is quite able to cope. If the intake of fat becomes too large, as for example after a rich meal, excess fat is stored as globules in the liver tissue. If these conditions are exceptional, no harm is done, but permanent excess leads to liver damage or cirrhosis.

The best-known product of the liver is bile, which passes into the duodenum via the gall-bladder. Bile acts in two ways: it neutralizes the acidity of the stomach so that the pancreatic enzymes may work on the partially digested food, and it also acts as a detergent or emulsifier in the digestion of lipids. Although fatty particles will never become soluble in water, if they are broken down into small enough globules an emulsion results. The necessary enzymes are able to get near enough by approaching the edges of the globule and so break down even essentially insoluble particles. So bile acts rather like the detergents—which give good wetting—and the emulsifying agents used in, say, the making of a haircream in which water and oil have to be brought into intimate contact.

There are several ways in which the liver acts as a scavenger to the end products of various degradative processes in the body. For instance, dead blood cells find their way to the liver, where they are broken down by the so-called Kupffer cells. The end products of this process are a variety of dark pigments which pass to the duodenum in the bile and colour the faeces. There are also a great variety of enzymes which act by breaking down other end products. Degradation of proteins and amino acids in the system can result in the formation of ammonia, which if left free would have disastrous consequences. The liver, however, transfers the free ammonia into the urea cycle, the end product of which, urea, is passed through to the kidneys to be finally eliminated in urine. Similar though not necessarily as complicated cycles also take care of a number of drugs, or drug-like substances which the body may either ingest or manufacture itself. In general, the liver renders these materials harmless by making an inactive derivative which can be eliminated from the system. One way to observe that the liver is not working properly is to note that harmful substances, which under normal conditions would have been transferred into an inactive variety, begin to appear in the urine.

THE KIDNEY

Of all organs the kidney resembles most those vast, unseen departments in cities and factories that, unsung, unheroic, and unromantic, see to it that life can continue in a reasonable manner: that water is drinkable, that sewage is rendered harmless, that dustbins are emptied every week. In physiological terms the kidney is a filtering and pumping unit which removes impurities from the system and sees to it that the essential water balance is maintained.

This latter function is a logical consequence of our condition as land animals whose bodies consist mainly of water. If the water content of our system is allowed to decrease too much, we die. But at the same time it is necessary to flush out the harmful degradation products. Thus water must be used, but used in the most economical manner.

One of the best illustrations of how such requirements are fulfilled is to observe what different organisms do with protein wastes. However protein molecules are broken down, the end result is a large quantity of nitrogenous waste. Theoretically at least, there are a number of ways the system could cope: for instance, it could transform it into nitrogen gas, and, by analogy with carbon dioxide, get rid of it through the lungs. This would, however, be an energetically difficult thing to do, and in addition nitrogen is so insoluble in water solutions such as the blood that enormous transport problems would have to be overcome. So we find that with the exception of some bacteria, nitrogen gas is not produced by animals.

Another option is to convert nitrogen into ammonia (NH_3) and flush it out with water. This is a much more manageable idea, but there is a snag—ammonia in appreciable concentrations is highly poisonous. So this method can only be used if the ammonia concentration can always be kept low—and we find that this is indeed the method adopted by a number of sea animals, where the ocean currents can be expected to keep the total ammonia concentration down.

Once the animal exists on dry land, it also has to overcome the problem that water for these operations is relatively scarce, and must be kept at a stable level within the organism, otherwise its metabolic processes cannot operate. It must also make sure that no waste product either inside the body or in its immediate vicinity builds up to dangerous concentrations.

Birds have an additional problem: their embryos develop in closed eggs so that their waste products stay with them until they are hatched. The method of discarding nitrogenous wastes in birds has therefore evolved on the lines of solid disposal. Birds excrete uric acid, which is insoluble in water. Uric acid crystals are not dangerous even if they share the same egg with the unhatched chick, simply because they cannot return into circulation.

In mammals the water situation is somewhat more liberal. They also possess the mechanism of rearing their young in their own portable incubator, which is connected up for waste disposal. We find that mammals, together with a number of other animals, have evolved the production of urea and its flushing away through the kidneys.

In order to see how this and similar mechanisms work, let us consider the way the kidney functions. The waste products arrive at the kidney in the blood, being pumped through the renal artery. The artery breaks up into a very large number of small

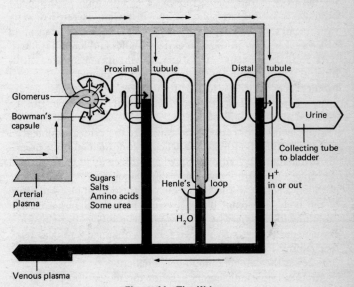

Figure 11   The Kidney

capillaries which form a ball-like object, the glomerulus. The glomerulus is enclosed in a system of small tubes, the nephron, called at this point Bowman's capsule. From Bowman's capsule the fluid is led into a convoluted tube, which itself consists of the proximal and distal parts. Between these sections lies the long and narrow Henle's loop. At the end of the distal tubule the collecting tubules lead to the renal pelvis and, through the ureter to the urinary bladder.

The general and oversimplified description of the kidney's work is reasonably clear. It is only when details are considered that it becomes increasingly complicated and still not completely understood. Basically, at any one time the kidney contains one quarter of the total blood supply undergoing filtration and reabsorption. Filtration takes place in the nephrons at the rate of 125 millilitres per minute. This process takes out the waste products such as urea, but leaves in the blood stream proteins and other large molecules unable to penetrate the nephron sieve. Filtration at this point produces a very dilute solution and if it were excreted directly would lead to an impossible demand for water. So a reabsorption process takes place in the convoluted tubules and loop that hands back 70 to 80 per cent of the water, leaving a concentrated ultrafiltrate. The ultrafiltrate, containing urine formed in the distal tubules, goes its way into the storage area in the urine bladder from whence, as our textbook charmingly puts it, it can be excreted at one's convenience.

Complexities arise when one considers the central problem in kidney function: how to keep a delicately adjusted multiple pump-filter unit running continuously when a diversity of loads vary all the time. One of the most important factors is the total volume of water. In other words, irrespective of the overall blood pressure of the body, the renal blood pressure must be above a certain minimum level, otherwise there would be insufficient driving force to make the kidneys work properly. This is obtained through a relay mechanism that constricts the renal artery and therefore increases pressure as the volume of fluid gets smaller.

Another important factor, perhaps the most important in the kidney's role in relation to the total system, is the maintenance of the constancy of internal environment—the so-called homeostasis. This involves not only the maintenance of a constant volume of fluid in the system—in terms of blood plasma and intercellular fluid—but also the maintenance of this fluid at a constant composition. For instance, protein breakdown products may contain sulphur that ultimately appears as sulphuric acid. This has to be taken away to conserve the correct acidity of the body as a whole. There is a complex hormone-controlled mechanism built into the kidney function which permits excess acid to leave and at the same time preserves electrical neutrality across the membranes. If this mechanism breaks down, the body begins to suffer from acidosis, a build-up of acid.

Similar necessities also apply to a range of dissolved molecules, which together determine the osmotic pressure of body fluids. Without defining osmotic pressure at this stage, we can regard it as a measure of the total content of solids in the solution. If, for instance, a large number of these small molecules are filtered out, a similarly large amount of water will be needed for their reabsorption.

Another example of the latter function is the case of the diabetic whose insulin supply has failed. Insulin is used in the breakdown of the sugar, glucose. If insulin is not present in sufficient quantities, the kidney attempts to take out of the system as much glucose as possible and some of its breakdown products, the ketone bodies. A limit is eventually reached when it can no longer cope and the accumulation of ketone bodies increases body acidity until the patient dies.

The general function of the kidneys is therefore the maintenance of the correct volume and composition of the body fluids. This it carries out by filtering and pumping to waste unusable degradation products and by controlling the rate of filtering and the reabsorption of still usable material.

CHAPTER 4

proteins and enzymes

'A substance in plants and animals, without doubt the most important of all known substances in living matter . . . this material has been named protein.' So started the scientific career of proteins 130 years ago, when they were first recognized by Mulder, a Dutch chemist. Between this simple and inaccurate observation and the results of today's sophisticated team efforts in all the major countries of the world stretches a history of scientific endeavour which can take its place among the greatest human achievements.

Some of the agencies contributing to the fabric and the functions of all living matter are very large molecules indeed. Proteins are members of this class of compounds. Their size ranges from about 13 000 to several million times that of the hydrogen atom, or 1000 to 100 000 times the size of the water molecule. As we shall see (cf. page 74) they are all composed of amino acid units. This immense range of size is accompanied by a parallel diversity of function.

The most important role of proteins is that they provide catalysts, enzymes, for biochemical reactions. All enzymes are proteins, and enzymes can be considered as primary determinants of cell function and therefore of the nature of the organism. Although one can argue that the fundamental characteristics of the cell are encoded in its genetic make-up, yet the expression of these is displayed through the production of specific enzymes. The characteristics of a cell can therefore be described in terms of the enzymes it manufactures.

All enzymes are proteins but not all proteins are enzymes. All that is said about proteins is generally also valid for enzymes, although because of their great importance the latter will be discussed in a separate section. From the architectural point of view, all proteins are generally considered under two headings: according to their shape they are described as fibrous or globular species.

FIBROUS PROTEINS

Fibrous proteins tend to be long, thin molecules that look like fibres. They form sheets, as in nails and the skin, or bundles, as in hair and muscles. Proteins associated with biological activity tend to be globular. Long, thin proteins are much less soluble than short, fat ones of the same weight. This property neatly fits in with their respective tasks: structural proteins remain in one position normally, while functional proteins have to be relatively mobile and active in solution.

Despite the seeming reasonableness of such a division, it has its dangers. For example, when we suggest that fibrous proteins have a structural role, we must not imply that this role is purely structural. As we shall be arguing in subsequent chapters, large parts of biochemistry can be regarded as similar to surface chemistry: reactions take place on and are influenced by a surface. In biochemical systems, such surfaces are often formed by material containing proteins, as for example in membranes. It is becoming clear that such surfaces have a significant influence on the reactions that take place upon or through them, and evidence is also becoming available that surfaces may take an active, dynamic part in these reactions. We should bear in mind, therefore, that even a protein in a structural situation may be a very active participant, rather than a passive spectator. Polymeric enzymes, for example, can form part of a membrane structure. It has also been argued that in a number of systems the deformations of a membrane provide the changing environment for subsequent reactions between reacting species, leading to different products. We shall also present the argument that considers enzymes to consist of one or a few catalytic areas kept in position by a structural framework; in such a case the protein frame can be described as making a structural contribution.

A striking demonstration of the importance of fibrous proteins occurs in muscle fibres. The contraction of muscles, which enables them to do such work as moving limbs, or in the case of the heart walls to pump blood, depends on the interaction of two different proteins called actin and myosin. This will be explained in detail in Chapter 7.

Proteins also provide a matrix for other tissues of the body. A parallel for this function exists in man-made materials; for instance in polyester-glass fibre boats, sheets of glass fibres form a matrix that is impregnated with resin to form a tough, rugged construction.

Similarly, the connective tissue of the body is made of a special protein called collagen, providing a matrix in which the organs of the body are supported. Collagen, made by special cells called fibroblasts, also occurs in tendons. Since it is very tough, high collagen content in meat leads to a tough joint. On boiling, collagen changes to gelatin, which is soft.

Proteins are also found among the substances composing the cell walls and may be involved in their function. These 100 Å (one millionth of a centimetre) thin membranes are concerned with regulating the composition of the cell by controlling the entry and outflow of material, but despite their great importance, neither their exact composition nor their mode of function is as yet completely understood. In a number of instances, proteins are attached to some other material and this partnership acts as a unit. When this happens, the unit is called a conjugated protein. Those in the cell membrane are one example: there the protein molecule is attached to fat-like material (phospholipid). Other instances of conjugated proteins occur among the functional activities of these materials.

FUNCTIONAL PROTEINS

Enzymes and some hormones consist of proteins, either totally or in part. Enzymes are the highly specialized catalysts which ensure that all reactions in the body necessary for the maintenance of life go ahead at the correct speed, and they are so important that we shall discuss them in great detail later.

Hormones are another group of substances involved in the promotion and control of metabolic processes, and a number of them are proteins or polypeptides—small proteins. It is characteristic of hormones that they are carried in the blood stream and act at a distance from the place where they are produced. A current theory suggests that they act by interfering with the control mechanism of individual cells.

An example of these biological journeymen is insulin, produced in the pancreas by a group of cells called the islets of Langerhans, named after the Dutch anatomist who first described them. Insulin is involved in the metabolism of carbohydrate in the body (cf. Chapter 6) and its absence can cause diabetes. Conversely, some types of diabetes may be corrected by injections of insulin.

One of the best-known functions of particular proteins is that of carrying oxygen. The carrier in the red blood corpuscles, which transports oxygen from the lungs to the tissues, is a protein called haemoglobin. To this entity is attached an additional non-protein part, called the haem group, a specialist in oxygen transport. With the help of a constituent iron atom it can take up, carry, and unload oxygen very rapidly. Haemoglobin and similar conjugated proteins are of great antiquity on the evolutionary scale, and even some of the most rudimentary animals contain structures similar to it, although they are not necessarily oxygen carriers.

If we think of haemoglobin as the wholesale distributor of oxygen, carrying its merchandise to all parts of the body, one of the retailers is another conjugated protein, called myoglobin, which serves as the oxygen carrier in muscle. Myoglobin is smaller than haemoglobin and has the distinction of being one of the very few protein molecules whose structure has been completely elucidated.

Oxygen collected by the lungs is made to combine with hydrogen from the breakdown of foodstuffs within the cells of the body. This final merger proceeds through a complicated series of steps which produce, among others, a molecule called ATP, adenosine triphosphate. In its turn ATP is required for the synthesis from the breakdown products of food of materials necessary for the survival of the organism.

The carefully programmed oxygen-hydrogen reaction systems

within the cells take place through the intervention of cyto-chromes, a type of conjugated protein located in arrays on mitochondria, the specialized organelles responsible for cell respiration.

If one separates from blood all suspended matter such as blood cells, one obtains a colourless liquid called blood plasma. This liquid is full of proteins; one of their main functions is to prevent water in the blood from escaping into the neighbouring tissue, in short, to keep the osmotic pressure steady. Blood plasma also acts as a carrier for a range of materials, hormones, enzymes, and antibodies among them.

Other plasma proteins include the globular transferrins, carriers of non-haem iron, the gamma-globulin antibodies, the albumens—familiar to anyone who has ever boiled an egg and noticed it become hard as a result of albumen polymerization. Plasma proteins also include a number of enzymes.

The oxygen-carrying capacity of the blood sometimes becomes faulty. In sickle-cell anaemia a very small defect occurs in the protein part of the haemoglobin. Two out of its 574 units are of the wrong type, and the result of this seemingly unimportant change is an incapacitating disease. Yet by an irony of nature, the affliction has its compensations.

It was found that sickle-cell anaemia is more widespread in the malarial regions of Africa than anywhere else and that while the malarial parasite carried by mosquitoes destroys the red corpuscles of normal blood, it has far less of a harmful effect on the blood of those who suffer from sickle-cell disease. We have, therefore, a paradoxical situation, in that the disease is advantageous for sur-vival as far as malaria is concerned, yet at the same time it has destructive effects of its own.

ANTIBODIES

We now come to one of the most fascinating areas of protein function: their role in the body's fight against disease.

When the body is attacked by bacteria or viruses, it has two principal ways of fighting back. White blood cells (leucocytes) can

move to the point of attack to engulf and destroy the foreign
agents. Secondly, a new and very specific protein, called an
antibody, can be produced which is transported in the blood stream
to neutralize the invaders. The part of the attacking organism
responsible for stimulating antibody production is called an
antigen.

The antigen may itself be a protein, or a sugar-like molecule
called a polysaccharide, associated with protein from the outer
structure of a bacterium. Alternatively, it may be a toxic material
—a toxin—produced by the invader. In every case, the antigen

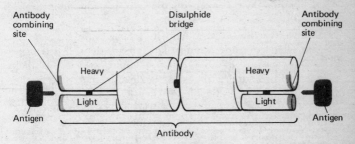

Figure 12   Antibodies

Antibodies are made of two heavy and two light protein chains linked together by
disulphide bridges. Antigens are captured by the antibody at the 'combining sites'
at both ends. The antigen is thereby neutralized.

is a large molecule, its specific action being to stimulate the production of the complementary antibody. We are not certain how
the antibody neutralizes the attacking agents; it may force the
attackers to change their outer structure, so that the task of the
white blood cells becomes easier. Alternatively, antibodies may
cause precipitation of the attacking cells or the toxins produced
by them.

The mechanism by which antigens cause antibody production
is at present the subject of intensive research and has yet to be
satisfactorily explained. A most important question is, how does
the body know which antibody to produce in response to an
invading organism when it has had no experience of it previously?
Despite our ignorance of the mechanism, the fact of antigen-

antibody reactions provides us with a powerful weapon against bacteria and virus-borne disease.

When a bacterium or virus attacks, the body needs time to build up a sufficient supply of antibodies for counter-attack. This time-lag can be made to disappear if one can arrange for the right antibodies to be already present in the blood stream. Antigens to stimulate antibody production are present in dead or non-virulent organisms, and in *vaccination* these are introduced into the body, which then produces the appropriate antibodies without falling prey to the illness. Thus, smallpox can be prevented by giving an injection of a mild virus closely related to the more harmful variety of smallpox.

Late in the eighteenth century Dr. Jenner, a physician practising in Gloucestershire, found that dairymaids who had been exposed to cowpox did not catch smallpox, which was then endemic in England, or if they did, the disease occurred in a mild form. From this observation Jenner developed the idea of vaccination, which soon became popular. Indeed, a vaccination mark came to be as fashionable as it was useful.

Today, a hundred and fifty years after Jenner's time, exactly the same principles are used in vaccinations against polio. When killed or attenuated (weakened) polio viruses are injected, the body produces the antibodies necessary to guard against future invasion by the full-strength virus.

Rabies and cholera can also be prevented by vaccination; the story of how Pasteur developed the rabies vaccine marks one of the milestones in the history of preventive medicine.

'Three individuals from Alsace unexpectedly presented themselves at my laboratory on Monday, the 6th of last July,' wrote Pasteur in 1885. One of them, a nine-year-old boy, Joseph Meister, who had been bitten fourteen times by a dog suffering from rabies, had been brought to Paris as a last resort. At the time Pasteur had developed a method of vaccinating dogs against rabies by injecting them with attenuated viruses. To produce the virus, an extract of the spinal cord of a dog suffering from the disease was first injected into a rabbit. The rabbit promptly got rabies and in turn its spinal cord was injected into a second rabbit and the process continued until about the twentieth animal had been injected. By that time Pasteur was obtaining standard incubation periods of seven days. He then suspended in air for several days the spinal cord taken from such an infected animal, put some of it in sterilized broth, and the vaccine was ready.

Joseph Meister presented the crucial test—although dogs could be saved, there was no certainty that humans would respond to this novel treatment. But without vaccination, the boy would certainly have died, so after a great deal of anxious hesitation Pasteur decided on a course of injections. Joseph Meister lived. Some months afterwards Pasteur wrote: 'His state of health leaves nothing to be desired.'

While vaccination stimulates the body's antibody production by providing dead or attenuated bacteria or viruses, *immunization* involves the injection of modified toxins—the harmful products of the attacking agents. Toxins carry their own antigen and therefore can call forth a good supply of antibodies. Diphtheria and tetanus can be prevented by immunization.

BLOOD GROUPS

Immunity to disease provided by antibodies in the blood serum is not inherited. However, an important group of these proteins is passed on by inheritance: the antigens and antibodies of the blood, which are responsible for the characteristics used to classify blood into groups. There are four main blood groups, A, B, AB, and O, although a number of subsidiary ones are also known. Blood grouping recognizes the complicated fact that blood contains a double ration of proteins controlling its type.

There are two antigens, A and B, which can be located in the cell walls of the red blood corpuscles, both belonging to the sugar-like mucopolysaccharides. A given individual may possess antigen-A (A group), antigen-B (B group), both antigen-A and antigen-B (AB group), or neither (O group). In addition, antibodies to the A and B antigens may also be present in the blood serum. The effect of the antibodies on their respective antigens is to aggregate the red blood cells containing them—in other words to put them out of action by curdling the blood. An individual with antigen-A cannot therefore have antibody-A. Instead his blood serum will contain antibody-B. Similarly, antigen-B is accompanied by antibody-A. No antigens, as in group O, means that there are both A and B antibodies, while possession of both antigens, in group AB, must bring about the absence of both A and B antibodies.

Disastrous early experiences have shown that blood grouping is of primary importance in blood transfusions, when blood from two individuals is mixed. If the red blood corpuscles in the donor blood are destroyed by the antibodies in the receiver's blood serum, the whole purpose of the transfusion is lost. On the other hand, it does not greatly matter for a first transfusion if the antibodies in the donor blood attack the antigens of the receiver, as the added serum will only represent a small part of the extra blood supply.

O-type blood can be given to all groups, since it does not contain antigens. A-group blood is destroyed by antigens in both B and O blood sera, B-group blood by A and O sera, and AB blood by A, B, and O sera. If this sounds as complicated as it is, Table 1 may help to summarize all the permutations.

Table 1

Blood Group	Antigen	Antibody	Can Be Given to Groups	Can Receive from Groups
A	A	anti-B	A and AB	A and O
B	B	anti-A	B and AB	B and O
AB	A and B	none	AB	all groups
O	none	anti-A and anti-B	all groups	O

One should perhaps ask the question: What benefit is the existence of blood groups to the individual or to the species? And how does an individual come to possess anti-A antibodies, when it has never come into contact with antigen-A? The simple answer to these questions is that we do not yet know, although a great many research workers would give a great deal to find out.

ALLERGY

Even a response as crucial as the antibody–antigen reaction can still go wrong. The first injection of a particular antigen may produce little apparent response, but a second injection a few weeks later may result in a violent reaction. People who have suffered increasingly severe pains from wasp stings can vouch for the truth of this statement. The severity of the second reaction is due to an uncontrolled battle between the antibodies produced after the first injection and the antigens absorbed in the second dose. The phenomenon is known as *hypersensitization* and the reaction is called *anaphylaxis* (against protection).

People susceptible to anaphylaxis often suffer from allergies. Allergies range in gravity all down the scale, although fortunately in most cases they are not dangerous. Like hypersensitization, an allergy is a neurosis of the antibody reaction; the violence of the effect bears no relationship to the gravity of the cause. Where an allergic person is concerned, normally harmless proteins such as those contained in pollen or cat fur cause a violent reaction resulting in a range of symptoms all the way from a skin rash to anaphylactic shock. The reason for this is unknown, although it is possible that the antibodies are localized too near the surface of the cells and are therefore overexposed. Since one does not know exactly what is happening, medicine can only alleviate the symptoms by administering drugs which reduce the violence of the reaction and by counselling the patient to keep away from whatever causes his allergy. Recently, it was found that injections of graduated doses of offending protein will alleviate and sometimes cure an allergy—hay fever, for example.

Similar effects are more tragic in consequence when they interfere with the possible transplantation of tissue or even organs from one person to another. The mitigation of such rejection reactions is the subject of much research, and only recently was recognized by a Nobel Prize awarded to Professor Medawar, then working at University College, London.

In their most extreme form, antibody reactions can also develop between different parts of the body. In the so-called auto-immune diseases, one part of the organism starts to make antigens. The rest of the body reacts by producing antibodies to destroy the offending tissue. In this chimereal civil war only the patient suffers. While the disease can be coped with, it is perhaps more important to find out why this type of reaction happens, and what agency prevents such tissue reactions in the normal organism.

CONTROL FUNCTION OF PROTEINS

Proteins are present within the regulatory mechanism of the cell. These are called histones and are associated in some way with the control of the cell processes and its replication (cf. Chapter 6). It is possible that they have some role in organizing the cell's activities, although there is as yet no clear idea of how histones exert their influence on the expression of genetic information.

Finally, we should consider briefly certain ideas which suggest that proteins have a major function in controlling the development and differentiation of cells. Once again we must bear in mind that these hypotheses by no means are proven and may well turn out to be false.

Every organism starts life as a single cell which by division produces more cells. As we have seen, the daughter cells in turn continue to divide, until the complete organism is developed. At an early stage, all these cells are identical. However, after a few hundred divisions, a change takes place and the cells produced become different both in size and in function. Differentiation has taken place, so that in the fully grown individual highly specialized cells perform different, accurately specified jobs. It is possible that such a differentiation process takes place under the control of proteins. Specific proteins produced at different stages of development have been isolated and have been observed to cause enhanced growth and differentiation in selected groups of cells.

CORRELATION BETWEEN FUNCTION AND STRUCTURE OF PROTEINS

Evidently in order to fulfil all these different functions, proteins must be specialized. In the same way that identical types of building units—bricks, wood, and cement—can provide the components for houses of different size, shape, and use, the characteristics of proteins are the consequences of the number and arrangement of their constituent units. All these belong to one group of substances called amino acids (cf. page 74). The variation in the number of units and their arrangement gives rise to an extraordinarily wide range of unique proteins.

How can one account for the varied roles and functions that protein molecules carry out in the living organism? The underlying principle must be that where a protein is, and what it will do, are determined by its constitution: what the protein consists of and how its components are put together. Once these facts are known, we can make reasonable guesses about the functions of the various parts of the molecule, and test these hypotheses by experiment.

In the early days it was thought that there was only a single type of protein molecule, but we now know that there are many thousands of different sorts of proteins in the human body alone. This figure suggests the order of magnitude of the difficulties research has to overcome.

Like Mrs. Beeton's hare, the protein molecule first has to be caught. Even when netted, it is the scientist's task to prove that his catch is truly a protein, and preferably the one he has been looking for.

SEPARATION OF PROTEINS

There are a number of standard methods which are suitable for separating and identifying the usual range of small molecules. Unfortunately, hardly any of them can be used in protein chemistry, principally because protein molecules are very large and the differences between types are rather small.

A further difficulty lies in finding a yardstick by which the progress of protein separation can be measured. Proteins are, in general, very similar in their composition and therefore cannot be differentiated easily by physical or chemical tests, like the usual run of small molecules. There is, however, an interesting test that may be applied. Proteins frequently have highly specific biological functions and it may be possible to estimate the quantity of a particular protein species by noting its biological activity. This is possible in the case of enzymes and also in the case of proteins or polypeptides which cause specific changes in whole animals or isolated tissues.

After purifying the protein, the amount of biological activity of its solution is measured. As the protein becomes more and more pure, the biological activity in terms of protein concentration will rise to a maximum, which is characteristic of the completely purified protein. If we are faced with a protein which does not have a measurable characteristic, the problem becomes much greater. This is probably the reason why most proteins that have so far been isolated are those which show obvious measurable properties.

Let us therefore consider methods of separating proteins from extraneous matter and also from each other. A standard university course mentions six such techniques and associated variations, but for our present purposes we can consider them in two major groups.

One group of separations is based on the difference in solubility of different proteins in a given solvent. A solid will dissolve in a liquid because any given molecule of the solid will be more attracted to the molecules of the liquid than to those of the solid. If we reduce the available numbers of liquid molecules, their attraction will decrease and eventually a molecule which was originally joined with others to form a solid will again find it more profitable to revert to its previous state. When this happens, the solid will precipitate from the liquid. Because the attractive forces between their molecules are different, different protein molecules will precipitate at different stages in varying liquid environments.

We can vary the number of available liquid molecules in a given solution in a number of ways. We can, for example, add a salt such as ammonium sulphate, which effectively mops up a number of liquid molecules. We can also change the forces between the solid molecules by adding an organic solvent which alters the electrical properties of the solution. The same result can also be achieved by altering the acidity of the solution.

These techniques have two principal aims. First, by varying the conditions in the solution, we can separate different protein molecules. Secondly, uniform behaviour in going into solution and coming out of it is a good confirmation that we have indeed obtained a pure protein.

A widely used method is chromatography, which also depends on slight differences in characteristics between different protein molecules in a given environment. Here we are making use of the varying powers of adsorption of proteins on various chemical substances. We shall have more to say about chromatography further on (cf. page 81).

Another group of techniques for separating proteins depends on the effect of applied force. Here, we have to choose a force that is suitable for the size of the molecule—the usual ones are electrical or gravitational. A limitation of these methods is that they can only be applied at a late stage in purification, when there are no more than two or three types of proteins present.

Electrophoresis is the name given to the method in which electrical forces are used. An electric current is passed through a solution of proteins and the molecules move under the influence of the current, each with its own specific speed, depending on the number and type of electric charges each carries.

Gravity is called in as a protein separator when the ultra-centrifuge is used. This instrument is basically a mechanized sling, although together with all its ancillary equipment it may not leave much change from £5000. A solution of proteins is put into a tube which is then whirled round at speeds approaching 60 000 revolutions per minute. In consequence, there is a force of approximately 100 000 times normal gravity acting on the tube and its contents. The force causes the molecules to move along, the extent of the movement depending on the mass of each individual molecule. Thus the heavier molecules move at a faster rate than the lighter ones. The position of a protein molecule after a certain time of centrifugation and its rate of progress under the gravitational forces are related to its mass.

In order to follow the migration of proteins during centrifugation, one can pass a beam of light through the centrifuge tubes at different time intervals. The light will be bent (refracted) in those regions of the tube containing protein, and the extent of refraction will be related to the amount of protein present. Regions of the tube containing pure solvent will let the light beam pass with very little refraction. Thus, it is possible to follow the migration of proteins during centrifugation, by optical means. The position and concentration of proteins in different parts of the tube can be represented on a graph as a series of peaks, the size of which are a measure of the amount of protein present. When only one peak is seen, one can reasonably assume that the solution contains only one major protein. Unfortunately, the optical system is not sensitive enough to show contaminating proteins if their concentration is less than about 5 per cent. Electrophoresis and ultracentrifugation can also be used to give an indication of protein purity. If our solution behaves as a single, homogenous entity, we can assume that only one protein is present.

Once we have obtained the protein molecule, we have to find out what it consists of. There are two answers to this question. In the first instance we can break down the molecule into its constituent elements and find out by classical methods what these are and in what proportions they are present. This analysis has been carried out on a number of proteins and it was always found that carbon, nitrogen, oxygen, and hydrogen, with perhaps some sulphur or phosphorus and an occasional metal, were the only constituents.

Although quite interesting, this fact by itself is not very helpful. There are no clues here which would allow us even to start to imagine the ways in which these elements might be put together. However, about 70 years ago Emil Fischer, one of the most brilliant chemists of the late nineteenth century, found that proteins can be described as chain-like structures, the units of the chain consisting of one type of substance only. These are the amino acids.

Amino acids are a most versatile group of species, distant cousins of the ammonia (NH_3) molecules. Their versatility arises from their ability to act rather like those small domestic electric drills that take a range of different attachments. The general formula for an amino acid and some of its variations are shown in Figure 13. First, the carboxyl group (COOH) and the amino group (NH₂) are suitable for forming strong bonds, the so-called peptide bonds. These are in evidence not only in proteins but also in a

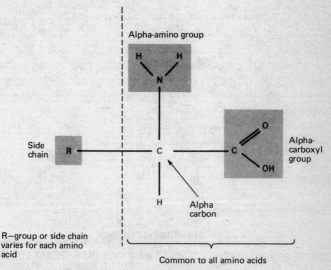

Figure 13 General Structure of an Amino Acid
Side chains can contain ring structures, chain-like structures, extra amino or carboxyl groups, or sulphur atoms.

man-made fibre such as nylon. But once the peptide link has formed, there are still groups left for reaction or ionization, depending on the nature of the R group or side chain. Even the NH group in the peptide bond has some residual attraction left that is displayed in its ability to take part in hydrogen bonding.

The electrophoretic separation of proteins, depending on

different charges, results from the ability of their amino acid units to ionize in different ways. Thus, one amino acid may contain an NH_3 group in the side chain; this can attract an additional hydrogen atom and can therefore show a positive charge. If the side chain contains an acidic COOH group, the amino acid will be negative. In this way the amino acids forming the protein, now called amino acid residues, can still be diverse even though they

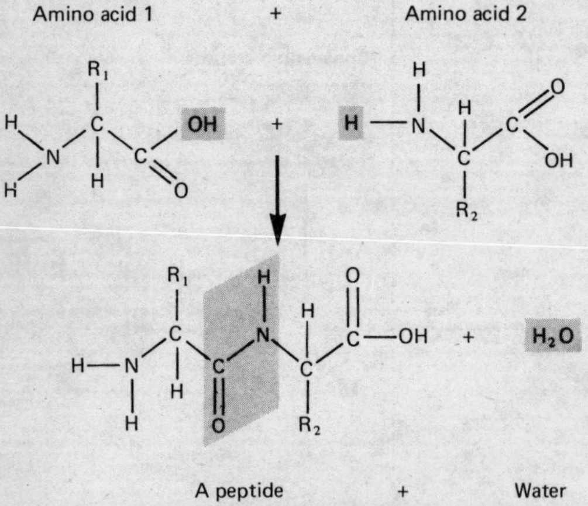

Figure 14 The Peptide Bond
In this diagram, two amino acids are joined by a peptide bond between the carboxyl group of one amino acid and the amino group of the other. This results in the removal of the elements of water. All the atoms participating in the peptide link are in the same plane in space. The side-chains (R_1 and R_2) stick out above or below this plane.

have a joint function. This diversity is essentially utilized in protein separation.

If a sufficient number of amino acids join together we can obtain very long chains. All present indications are that proteins consist of such polymer chains. The number of units in the chain varies widely: the hormone insulin, which is involved in the body's sugar metabolism, has 51 units. It is considered to be rather small fry.

At the other end of the scale we can consider ovalbumen, one of the principal ingredients in egg white. In this case 400 amino acid residues take part in the chain.

There are two very interesting points about these arrangements. First, despite the gigantic number of proteins in the human body, all of them consist of only about 20 different types of amino acids. Secondly, all these acids have the power to bend polarized light. We do not know the reasons for either of these facts. It is probable that they are connected in some way with our biological evolution, in that the extra information contained in the property of rotating light is in some way advantageous.

Once we have arrived at the idea of protein chains being formed from amino acid residue units, we can simplify a little by considering one structural problem after another. The first relates to the number of amino acids in a protein molecule and the sequence. This is referred to as the primary structure. Next, the chain is not simply stretched out, but tends to associate with similar chains, to form 'pleated sheets', or alternately, to coil up into a helix, usually described as an alpha helix. The extent to which this happens is determined in the secondary structure of the protein. In some proteins the alpha helix is not complete; instead it bends upon itself, for example in globular proteins. This is the tertiary structure. In others, for example in fibrous proteins, the coils coil around each other, again to give a tertiary structure. Finally, there can also be an association between coils or helices—this is called the quarternary structure. A main effect of all these twists and contortions is that units which would have been widely separated along a straight chain are brought close together and can interact.

We can start with the premise that we have been able to obtain a pure protein by one of the techniques we looked at previously. An idea of its size can be obtained by noting its behaviour in the ultracentrifuge. The next step is to find out which amino acids are present and the sequence in which they are joined.

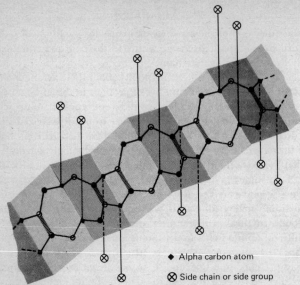

◆ Alpha carbon atom

⊗ Side chain or side group

(a) *Pleated sheet structure of some proteins*
Antiparallel pleated polypeptide chains are held together by 'hydrogen bonds' (||||||||||) formed between the neighbouring —CO group of one chain and the —NH group of another chain.

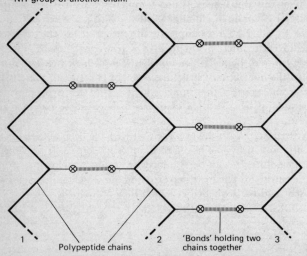

1   Polypeptide chains   2   'Bonds' holding two chains together   3

(b) Pleated sheets of hydrogen-bonded polypeptide chains can be joined in the other dimension through their side chains (⊗) and 'stacked up'. The forces that hold the sheets together are often due to weak interactions between side chains.

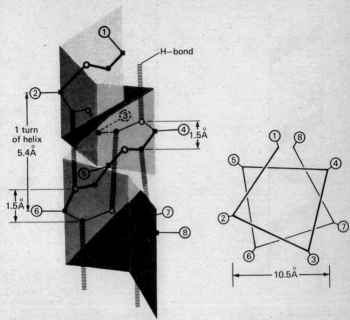

A side-view of the ∝-helix
Each surface represents the plane of a peptide group ●–○. Alpha C atoms ■ are on edges to which side groups, e.g. ②, are attached.

Top view of ∝-Helix
Planes are at angles of about 80° to each other

(c) *The α-Helix*
In this structure, the polypeptide chain is coiled on itself and held together partly by hydrogen bonding.

Figure 15   Secondary Structure of Proteins

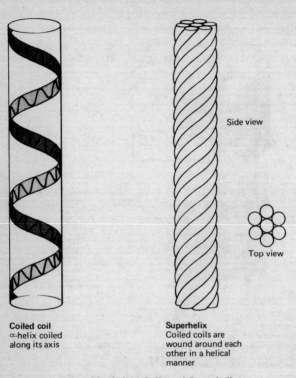

Coiled coil
α-helix coiled
along its axis

Side view

Top view

Superhelix
Coiled coils are
wound around each
other in a helical
manner

Figure 16   Coiled Coils and Superhelices
In hair keratin, several strands of α-helical keratin are wound around each other,
as in a cable, to form a superhelix of seven strands, six around one central one.

CHROMATOGRAPHY

The analysis of proteins depends to a large extent on a range of
techniques which come under the description of chromatography.
In principle these methods are very simple. In 1906 Michael
Tswett, a Russian botanist, was working on the constitution of
chlorophyll, a plant pigment. Twsett found that if the constituents
of chlorophyll were dissolved in petroleum ether and the solution

allowed to run down a tightly packed column of calcium carbonate, the pigments moved at different speeds. The different constituents could then be washed out one by one.

A development of this method is paper chromatography. If a drop of ink is placed on a blotting paper, concentric rings of different colours are formed as the constituents of the ink move from the centre at differing speeds. Instead of adding a large amount of solution to dry paper, we can put a small quantity of a solution containing a number of components on a very small area. One end of the paper is then dipped into a solvent. The solvent moves up the paper and when it reaches the spot it will carry along the constituents of the mixture at different speeds.

There are principally three different types of chromatography in use today: adsorption chromatography, where separation depends on materials being adsorbed to varying degrees on solid surfaces. In partition chromatography materials will dissolve to a different extent in stationary and moving solvents. In our example, the water in the blotting paper is the stationary phase, the solvent into which the paper is dipped is the moving phase.

A very important method is ion exchange chromatography, where an insoluble material containing ions on its surface exchanges them for ions in solution. The best-known examples of ion exchange materials are the columns used for water softening. Calcium and magnesium ions, which make water hard, are replaced by sodium ions.

Columns of ion exchange resins are also used for the separation of amino acid mixtures, and quite recently the whole process of separation, identification, and estimation of amino acids has been automated by three Americans, Spackman, Moore, and Stein. A protein is split into its constituents by boiling with a strong acid and the mixture run through an ion exchange column. As each amino acid comes off it is mixed with a reagent which makes it blue. The intensity of the colour, measured automatically, depends on the amount of each amino acid in the original mixture.

Basically, therefore, all that has to be done to find out what amino acids there are in a protein molecule is to break the peptide links and sort out the various components by putting the mixture through a chromatographic column or paper. Automatic analysis can greatly simplify the determination of the primary structure of a protein. Paper chromatography is very much more laborious. The paper must be sprayed with a reagent to colour amino acids, which are normally colourless. Each coloured spot representing a separated amino acid has to be cut out, and the colour extracted into solution before the amount of amino acid present can be determined.

Total amino acid analysis involves the breaking of every peptide bond in a protein molecule. Sequence analysis, a far more tricky operation, requires the breaking of fewer bonds to give di- and tripeptides—sequences of two or three amino acids, held together by peptide bonds—and finding the amino acid sequence in each fragment. In any peptide chain there is one amino acid which can be identified: this is the so-called N-terminal amino acid. It is the first in the chain and has therefore a free amino (NH_2-) group. A reagent (DNFB, dinitrofluorobenzene) will attach itself to this amino group so firmly that when the protein or peptide is broken into fragments, DNFB will still cling to the same amino acid. The presence of DNFB produces a yellow colour so that the marked amino acid can be separated from the others and identified by chromatography.

The whole operation of sequence analysis is the fitting together of an obscure jigsaw puzzle. Given that the protein molecule consists of a chain of amino acids, if all the tri- and dipeptide fragments are completely identified, it is at least possible theoretically to find the complete sequence by comparing overlapping portions.

First the protein is split into short, overlapping sequences of di- and tripeptides which are separated by chromatographic or other techniques. Each fragment is examined separately to determine its constitution: the examination includes the determination of the end amino acid by means of the DNFB technique.

The number of separate fragments necessary to identify a protein depends on its length—the longer the original molecule, the more sequences needed. For a very large chain, it is best to split it into two or more large fragments and to work on them separately. The initial splitting can be done by using special enzymes which attack certain bonds far more readily than others. Other enzymes attack amino acids at the ends of protein chains. When these are used, it is possible to get an indication of the amino-acid sequence in the separate fragments.

STRUCTURE OF INSULIN

An idea of the true difficulties that can be encountered in the determination of protein structure can be obtained from the efforts of Sanger in his work on the hormone insulin. The project took ten years, from 1944 to 1954, carried out in the days before automatic amino acid analysis, so that the laborious methods of paper chromatography had to be used.

Sanger showed that 51 amino acids make up the insulin molecule. After reacting insulin with DNFB, two different N-terminal amino acids were identified, showing that there are two chains, A and B. After separating the chains, one was seen to contain 30 amino acid units, the other 21. A further complication was met when it became clear that the two chains are connected in two places, through sulphur-containing pairs of amino acids called cysteine. The connecting links are called disulphide bridges. What had to be done was to separate the two chains, taking good care that neither was damaged in the process. Once this was achieved the separate chains had to be taken apart as gingerly and cleanly as possible, and the labelling methods used to discover the first amino acid in any sequence. Overlapping sequences then revealed the complete arrangement of all the amino acids in both chains. The work now had to take into account the connections between the chains, and to bear in mind that the disruption of these chains almost inevitably caused other changes which tended to mask the real state of affairs. After ten years, this work achieved its aim. It made possible the total test-tube synthesis of the insulin molecule, in 1967, by a group of mainland Chinese workers. In the summer of 1969, Dorothy Hodgkin announced in New York the X-ray structure of insulin.

THREE-DIMENSIONAL STRUCTURE OF PROTEINS

The protein molecule has been referred to as nature's most marvellous architectural creation. And we have indeed suggested that the role of proteins must be closely connected with their absolute arrangements in space. Up to and including the work of Sanger, we have been obtaining road maps (or partial ones) but not contour maps of the protein country. Clearly a great deal was still to be learned.

When one looks at the formula of a chemical compound as it is written on the page, one often tends to think of it as a flat, two-dimensional entity. This of course is not so. Not only have atoms definite volumes, but the links between atoms (which are used in forming molecules) have also accurately specified directions in space. The atomic volumes and bond directions are known in the case of the amino acids forming the sub-assemblies of protein molecules, and important results have been obtained by making models of these structures. Although a model on its own is no proof that one is on the right track, it does nevertheless suggest certain solutions which, with some luck, can be tested experimentally.

The making of protein models suggested some rather interesting possibilities. We have seen that all amino acids taking part in protein building are of one type, called the L-form. If we string together a number of such acids, we can make not only a straight chain, but also a structure similar to a spiral, called an alpha helix. This suggestion was first mooted by the American Nobel Prize winner Linus Pauling, and it appears that it is a very widely occurring protein construction.

Proof of this proposition was obtained by another experimental technique of very great importance in studying biologically vital molecules. This is called X-ray crystallography. X-rays are similar to the visible rays of light, but their wavelength is much shorter. We can assume that as a consequence of this, they are able to 'see' much smaller objects than visible light. Another way of looking at the same phenomenon is to say that objects of atomic dimensions will scatter, or diffract X-rays, and the pattern of the diffracted light gives an indication of the positions of atoms. The pioneering studies in X-ray diffractions were carried out in England by the late Sir William Bragg and his son Sir Lawrence, who was director of the Royal Institution. Their studies were made on relatively simple substances, but the same methods can also be applied to proteins.

X-ray methods were first applied to protein research in the early 1920s, when scientists at the Kaiser Wilhelm Institute at Berlin tried to take X-ray photographs of hair, horn, and similar substances. In England, W. T. Astbury, who has been called the father of molecular biology, carried out pioneering studies. The pictures had such a bad definition that they were considered, probably rightly, to be hopeless. Very considerable improvements in these techniques had to take place before they could be tried with any chance of success, and in fact about twenty years elapsed before they were again used in the investigation of protein structure.

An X-ray camera looks at the complete protein molecule. Theoretically at least, every one of the protein's constituent atoms can indicate its presence by a spot on the photographic plate, but the allocation of significance between the members of this small galaxy can cause great difficulties. So that when Linus Pauling and his co-workers began to use X-ray diffraction methods as a systematic means of studying the configuration of proteins, no conclusive results were obtained until computer programming could be brought in to help the task of analyzing results.

One of the most complete views of a protein structure available up to now has been obtained at Cambridge by Kendrew and Perutz. They investigated the protein myoglobin and haemoglobin, respectively, and based their findings on the analysis of 25 000 X-ray refractions (see Figure 17).

We can generalize these results, taken in conjunction with others. They have been obtained mostly in England and the United States by investigating protein segments and making polypeptide sequences artificially.

As the amino acids link up, they may form a helical structure—like a spiral staircase. We have good reason to believe that there

Figure 17

X-ray crystallography reveals the 3-D structure of proteins and other molecules. in oxygen transport and storage.

(a) The principle of X-ray diffraction

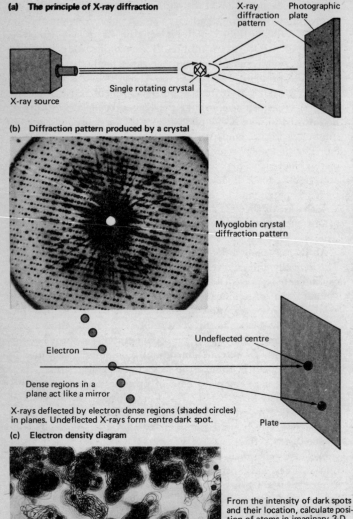

X-ray diffraction pattern

Photographic plate

X-ray source

Single rotating crystal

(b) Diffraction pattern produced by a crystal

Myoglobin crystal diffraction pattern

Electron

Undeflected centre

Dense regions in a plane act like a mirror

X-rays deflected by electron dense regions (shaded circles) in planes. Undeflected X-rays form centre dark spot.

Plate

(c) Electron density diagram

From the intensity of dark spots and their location, calculate position of atoms in imaginary 3-D 'unit cells'. Mathematical values defining place of atom in unit of space are 'translated' on to lucite sections as contour maps.
Hills = high density;
valleys = low density.

The Architecture of a Protein
How this is done is illustrated for the protein myoglobin, a muscle protein.

(d) Stacked lucite sheets

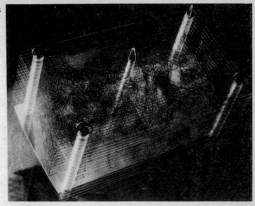

Principle of how a 3-D model of an imaginary protein would be extracted from lucite contour maps.

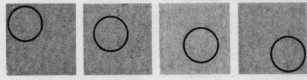

(1) Separate lucite sheets showing contours of simplified shape at different depths of 3-D model

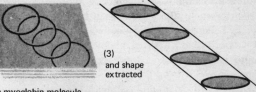

(2) Lucite sheets are stacked

(3) and shape extracted

(e) Model of a myoglobin molecule

are 3.6 amino acids for each turn of the helix. The helical form of the chain is the basic form, but we may have complications arising from the fact that some amino acid residues do not fit neatly into the helix pattern. Occasional kinks are formed, so that the chain may fold back on itself either partially, or along its entire length. Between such more or less paralleled portions of the chain we can also get bridges, formed by two sulphur atoms.

It is also possible for several chains to coil around each other, to form what have been called coiled coils. It seems probable that in situations where protein molecules fulfil a structural role, as in hair, the helices are twisted around each other, in the manner of a many-stranded cable (see Figure 16).

One may well ask what sort of forces can keep such an enormous display of atoms in such a precise, three-dimensional structure. In the first place, we have to go back to the basic fact about proteins: they are collections of amino acids. In their turn, amino acid molecules are rather tough entities: they do not distort easily. When these molecules link up, they cannot be compressed or extended to fill an available space. On the contrary, the spatial configuration of the resulting molecule has to fit in with the requirements of the amino acids.

There is also another set of cohesive forces. In the folded protein helix, hydrogen atoms bound to the nitrogen of one peptide bond come close to oxygen atoms bound to carbon on the fourth peptide bond along the chain. Now oxygen and hydrogen have a profound attraction for each other, even when both of them are formally satisfied by being a member of a molecule or group. This attractive power is described as a hydrogen bond, and in those proteins where its length can be measured it works out at something under 3 Å long.

Although a hydrogen bond is much weaker than a conventional bond, which can be found between atoms in a molecule, it does represent a measurable force. In the case of proteins, we must bear in mind that there are a very large number of opportunities for hydrogen bonds to form. Although each individual one is very weak, on the teamwork principle the totality of hydrogen bonds represents a very considerable force.

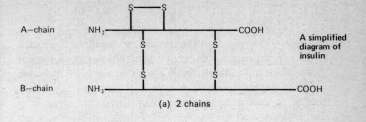

A simplified diagram of insulin

(a) 2 chains

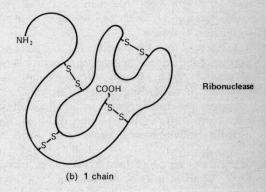

Ribonuclease

(b) 1 chain

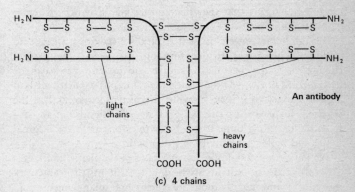

An antibody

(c) 4 chains

Figure 18 Disulphide Bridges are Important Structural Determinants of Proteins

STRUCTURAL VARIATIONS IN PROTEINS WITH DIFFERENT FUNCTIONS

Up to now we have followed the researches leading to some idea about the general constitution of proteins. The next two obvious questions can be put in these terms: given that a protein molecule is such an enormously complicated entity that we are as yet very far from being able to synthesize it (with the recent exception of insulin), how is it that the smallest cell of the humblest bacterium can do it with ease? Although there is as yet no completely satisfactory answer, we shall be looking at some promising theories in the next chapters.

The second question concerns the way proteins function. What portion of the protein molecule enables it to be a building block, an enzyme, or an antibody? Yet again, we must admit to our ignorance, but knowledge is slowly becoming available.

Certain guesses have, however, the chance of being correct. We have started this chapter by noting the simple results of observation, that proteins come in two different shapes. Some are fibrous and usually insoluble in water. Building-block types usually come in this form: keratin of the skin, silk fibroin, myosin of the muscle. Other proteins are globular and soluble: enzymes, hormones, serum proteins. It is usually assumed that most fibrous proteins are not coiled, i.e. that they have no tertiary structure, whereas globular ones do.

If a globular protein is heated or dissolved in a weak acid solution, a process called denaturation takes place. The protein changes its shape, becomes insoluble and in some cases can be described as fibrous. At the same time it loses its capacity for biological activity. If it is an enzyme, it will no longer catalyse the reactions it speeded up previously. If denaturation is not allowed to go on for a long time, it is possible to restore both the original shape and biological activity of the protein.

We have come full circle. Observations of the shape and large-scale properties of proteins are beginning to fit into the framework resulting from investigations of their molecular constitution. It seems reasonable to argue that denaturation means the disorganization of the globular protein, accompanied by the breaking of the hydrogen bonds. If this argument is correct, we must assume that hydrogen bonding has an intimate connection with the biological characteristics of the protein molecule. The key to these activities is hidden between the folds of the helix, the subtle relations between those intricate groups arranged in space, the architecture of 'the most important of all known substances.'

ENZYMES

The history of enzymes is a sub-plot in the history of proteins, with perhaps one important twist of its own. This was their role in discrediting the vitalist theory that asserted that although living things are made up from non-living chemicals, yet the whole was not only greater than the sum of its parts, but also capable of performing processes beyond the ken of non-living matter. Superficially, this is a most attractive theory; it has a quality of common sense difficult to argue against without seemingly resorting to sophistry. Indeed, we are beginning to say rather similar things even today; we assert that there are definite characteristics obtained from the property of size, that complicated feedback interactions do make the potential of the system greater than that of the sum of its components. What we are very careful not to do is to introduce any agency external to the materials and their interactions already in the system; we are playing the game very much by Occam's rule.

But before this essentially cybernetic or information theory approach, the vitalist theory did make a great deal of sense, even by the very simplifications it allowed. No wonder, therefore, that first-class scientists like Pasteur found nothing odd in its ideas. However, by the end of the nineteenth century the vitalist theory was crumbling under the attacks of the great mechanists, Liebig, Traube, and Claude Bernard. The end came in the last decade of the century through the work of the Buchner brothers, more as a spin-off than its main target.

The Buchner brothers were interested in obtaining protein injections for medical purposes, and yeast juice seemed a possible and a cheap raw material. To obtain their solutions they crushed the yeast cells in a hydraulic press and filtered off the cell-free yeast juice. The problem now was to preserve the juice; so, arguing along sensible cookery lines, they added sugar and waited for the result. The result was fermenting yeast juice and the start of a great deal of work on the mechanism of fermentation. But their results also showed that there was nothing in the healthy, all-natural yeast that could not also be included in its cell-free juices and analysed like all common or garden chemicals; hence the end of the vitalist theory.

The Buchners also found that if the yeast juice was boiled or dialysed no fermentation took place even on the addition of sugars. In dialysis the juice is placed in a porous container surrounded by water; the large molecules cannot pass through the sieve although the small ones can. By subsequent elaboration of such experiments it was realized by the 1920s that for enzyme action, not only the enzymes themselves are necessary, but also some small molecules called coenzymes, a number of which are vitamin derivatives. By 1926 the first enzyme was crystallized by Sumner in the United States. This was urease, catalysing the breakdown of urea into ammonia and carbon dioxide. Since that time over 1000 enzymes have been crystallized and enzymes have become not only of great biological but also of great industrial importance.

CATALYSIS AND SPECIFICITY

Sooner or later every book or article on biology or biochemistry comes out with some form of the statement: enzymes are the important, or vital, specific catalysts operating in living matter. We have looked briefly at the phenomenon of catalysis *per se*, and we shall also be considering aspects of this catalytic function from the point of view of control mechanism (cf. page 182). At this point it would seem sensible, therefore, to look at why enzymes are such good catalysts. In other words, how do enzymes gain the specificity they possess?

We should perhaps recall what a catalyst is and is not. A catalyst is not magic ingredient X which makes a reaction go when it would not do so otherwise. What it does do is to speed up a reaction. It will not change the position of equilibrium; it will not make the reaction go further.

Most catalysts, enzymes included, reappear unchanged at the end of the reaction they took part in, but not all. Unchangeability is no criterion of catalytic properties, whatever old-fashioned textbooks may say.

The catalytic effect of enzymes is powerful; in some cases about half a million reactant molecules are transformed per minute per molecule of enzyme, sometimes up to ten million. Thus it is just as well that such remarkable catalysts should be used for virtually all chemical reactions taking place in a living system. Even a bacterium has about 1000 enzymes, and as we pass up the evolutionary scale their number increases considerably.

Enzymes, in common with all catalysts, have a number of limitations. They are, for example, best at certain temperatures and degrees of acidity and, not surprisingly, for most enzymes operating in animal tissues this optimum point turns out to be about body temperature and near neutrality. The most important point about enzymes is their specificity; they will not catalyse just any reactant or reaction. Some will only affect the making or breaking of certain bonds, some can discriminate between different optically active forms of a molecule. Some, of course, are far less specific.

One of the most interesting puzzles about enzymes concerns the question of the mechanism or mechanisms of their specificity. There are two main answers that have to be given. Some explanation that is amenable to experimental verification must be provided as to why enzymes are indeed specific. But in addition the model must also be in agreement with the kinetic data available on the catalytic action of enzymes—these we shall look at when considering the role enzymes play in the coarse and fine control of cell chemical processes (cf. page 122). The kinetic effect of enzymes can best be illustrated by a graph (Figure 19).

The graph only expresses mathematically a truism: without having anything to react with, the enzyme will not show any activity. Therefore, assuming a ready supply of enzyme, the apparent activity will increase with the availability of something to react with, known as the substrate. However, as the substrate supply increases, a point will be reached when all available enzyme molecules will be fully employed; this is the point of

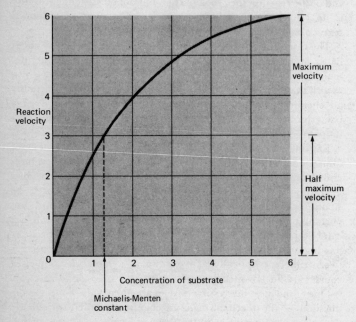

(The numbers above are arbitrary units)

Figure 19   Kinetics of Enzyme Action

maximum velocity of reaction for a given enzyme concentration with a given substrate. Obviously this point will vary according to the enzyme and the substrate, and is a measure of the detailed mechanism of enzyme action on the particular substrate. We can also note the halfway point to this maximum velocity and measure the amount of substrate necessary to reach it; the amount of substrate required for half-maximal velocity is called the

Michaelis-Menten constant. It gives some measure of the affinity the enzyme and the substrate have for one another, but the relation will be a reverse one: if the enzyme-substrate complex is very readily formed (low Michaelis-Menten constant), only a smaller amount of substrate will be necessary to show the same total activity than if the complex were sluggishly made.

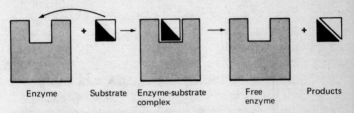

Enzyme   Substrate   Enzyme-substrate   Free   Products
complex   enzyme

Some poisons look like the substrates but cannot be broken down (transformed) by the enzyme.

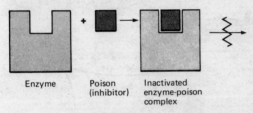

Enzyme   Poison   Inactivated
(inhibitor)   enzyme-poison
complex

Figure 20   Enzyme Poisons

We can therefore use these two measures, the maximal velocity and the Michaelis-Menten constant, to give some indication of the effectiveness of enzyme and substrate interaction, and their bearing on cell metabolism. They also show that, should some inhibitor or activator interfere with the course of the reaction by interfering with the attachment of the substrate to the enzyme, the difference will be noticeable in the value of the Michaelis-Menten constant although the maximal velocity might still remain the same.

The kinetic results require that we postulate some mechanism whereby the enzyme first makes contact with the molecule the reaction of which it promotes, the substrate, and that after a temporary association it then departs. In other words, there must be a more or less transient enzyme-substrate complex. If we then recall that all molecules have a very definite three-dimensional shape, the obvious supposition must be that enzymes are specific because their shape allows only one, or a few other molecules to come sufficiently near any individual enzyme for the substrate

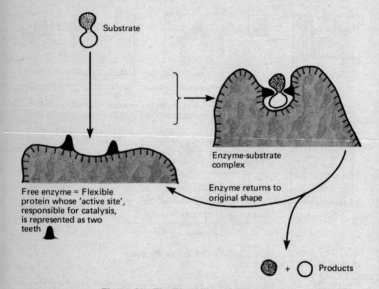

Figure 21 The Flexibility of Enzymes
The attachment of a substrate 'induces' a change of shape which brings the 'teeth' of the enzyme in the right place to 'cut' the substrate.

complex to be formed. This is the lock-and-key postulate. We assume that there is some type of geometrical matching between enzyme and substrate. Therefore, if one can provide the enzyme with a molecule other than its natural substrate but bearing a close resemblance to it, the enzyme should accept the fake substrate. We should therefore be able to poison the action of specific enzymes by providing specific, tailor-made chemicals for this purpose.

The lock-and-key theory, originally advanced by Emil Fischer, turns out to be correct, at least as far as it goes. The best example of how it works is the case of the sulpha drugs. Here the disease-causing bacteria rely on certain foodstuffs for growth; but if we present them with chemicals having a similar shape, the bacterial enzymes are fooled and incorporate our poison into their system instead of the raw materials required. In due course they die.

But the lock-and-key idea is essentially a static concept. Current work indicates that although enzyme specificity can still be thought of in terms of geometric fit, the arrangement itself is a result of dynamic action instead of the random coming together of two fixed shapes. The new idea, that of induced fit, suggests that the enzyme molecule is no more than a flexible framework for the inclusion of one or more active sites which carry on the re-quired work of catalysts. When the enzyme collides with the substrate molecule, it changes its shape, with the result that the active site is placed in a position where it can start to function.

We can further elaborate the idea of an enzyme undergoing a change of shape on collision with a substrate molecule and both of them in turn displaying a lock-and-key type of mechanism. Many enzymes are composed of a number of sub-units. Any change in the relationship of the sub-units will also affect the enzyme's capacity to undergo the postulated primary change of shape on collision with a substrate molecule. If we plot the kinetic effect of substrate concentration in the presence of excess amounts of enzyme, instead of obtaining the conventional curve showing increasing speed of the reaction with increasing substrate con-centration until the maximum is reached, we find a so-called sigmoid, or S-shaped, curve. Here, as we increase substrate con-centration, the speed of the reaction increases only slowly at the beginning. After a certain point, the increase in speed accelerates until it reaches a maximum. A given increase in substrate con-centration has therefore different effects, depending on how much substrate is already present.

Now it so happens that there is a very good model for this type of behaviour in another biochemically important material that is not an enzyme: this is haemoglobin, the oxygen carrier of the blood. The behaviour of haemoglobin is in some ways analogous to that of enzymes. Haemoglobin is a largish molecule, but the effective part is the so-called haem group containing an iron atom. It is this part that picks up an oxygen molecule and subsequently deposits it in the cells. Haemoglobin, which consists of four protein chains with a quarternary structure, reacts in a sigmoid curve if we relate the amount of oxygen carried to its concentration. Myoglobin, the oxygen carrier in muscle, composed of a single unit, shows a curve very similar to the conventional enzyme plots. Myoglobin is a much simpler molecule without quarternary structure.

Such results suggest that whenever the shape of the enzyme is complicated, a few substrate molecules have to induce some changes in shape. Once this has taken place, a change in general environment has occurred that facilitates a change in shape in subsequent enzyme molecules. This idea can be carried a stage further by considering molecules other than substrates as being capable of inducing such changes. Such molecules can be found; they are called allosteric effectors and act at allosteric sites. The idea of allosteric effectors came from studies of feedback inhibition, but it is possible that the same type of mechanism may also account for the activation of enzymes, especially by small, non-substrate molecules.

A further interesting feature brought forward by these ideas is their bearing on what are called iso-enzymes. These are groups of enzymes, all of which have the same active site, but different surrounding groups. Their differences are generally due to small variations in one or more sub-groups, and the distribution and forms of an iso-enzyme family in the same animal can vary from tissue to tissue. The main catalytic function of these enzymes is the same, but they show differences in minor properties such as in their responses to the different control mechanisms in a given cell. We can argue, therefore, that the existence of such iso-enzymes is a consequence of cell differentiation and hence that they are a reflection of the genetic pattern necessary to bring it about.

COENZYMES

Even in law, somebody has to guard the guardians, and in a similar fashion, it turns out that there are a range of smallish, heat-resistant molecules whose existence is necessary for enzymes to function correctly. One can therefore regard the former as sorts of catalysts for enzymes, although this definition would not meet the chemical or indeed the linguistic purist's approval. In a rough and ready manner one can say that coenzymes are essentially concerned with the products or the forerunners of the substrates the enzyme molecules have to operate on. Coenzymes of the type we shall meet in cell respiration (e.g. NAD) are carriers of protons and electrons. On this basis, although potentially the enzyme molecule is ready to act, it cannot do so if the products of the previous reaction are not carted away, or if in general carriers are not present for the input and outflow of material.

The interesting point is that most of these coenzymes first came under not the biochemists' 'microscope', but the physiologists' and the nutritionists'. Most coenzymes are either vitamins or derivatives of vitamins and their effects and existence were first pondered over by medical men. In actual fact, the very first time they came under examination, at any rate in the West, the motives were strictly utilitarian. In the nineteenth century, the British navy began to realize that on long sea voyages they were losing more men from scurvy than from enemy action, and through a series of accidents they came to realize that scurvy could be prevented by drinking lemon juice. With the rise of the Empire, lemon juice eventually became displaced by lime juice from the West Indies; this made good health and business sense and also provided the Americans with a convenient word of abuse for the British. A great deal before this time, in the thirteenth century, the Chinese had already realized that beri-beri could be avoided by a judicious choice of diet: eat rice with husks, as polished rice lacks vitamin B_1.

Disjointed observations of this type kept on accumulating, until by the turn of the century a reasonably large amount of knowledge was in existence about what we would today call deficiency diseases. The whole idea was very puzzling, for had not the giants of biology less than fifty years before showed conclusively that diseases are caused by germs? The idea therefore took some time to take root, and it was not before the second decade of this century that it became firmly established. By then it became obvious that merely feeding an animal—human or otherwise—the necessary calories, proteins, fats, and so forth was not enough. A great deal of experimentation was carried out by feeding rats with the pure elements of a diet: as a result the rats fell ill, developed haemorrhages and generally died. Unfortunately, the results were hardly consistent, probably because materials described as pure were far from being so and therefore provided the necessary ancillary materials.

However, in 1912, Hopkins in the United Kingdom conducted a series of classical, controlled experiments on young rats. One group was given a carefully purified basal diet which also contained pure casein, a milk protein. The other group was fed the same diet, but in addition was given small, controlled amounts of milk. This series of experiments showed conclusively that milk, even if only in very small quantities, was essential for healthy growth, and therefore must contain one or more vital factors.

Since then, the science and art of vitamins has grown apace. Today we divide vitamins into two groups: in general, the B vitamins are water-soluble and vitamins A and D (anti-rickets) are fat-soluble. Vitamin C, discovered in the 1930s by Szent-Gyorgyi in Hungary from extracts of paprika, is also water-soluble.

The main vitamins, what they are, what they do and what happens if the body does not acquire enough of them, are shown in Table 2.

The interesting point is that vitamins cannot be made by the organism at all, or only in very small amounts. From their physiological effects it is obvious that most of them are part of coenzymes and therefore essential to the healthy functioning of the system.

Table 2

VITAMINS

Name	Effects of Deficiency	Main Biochemical Function
Fat-soluble vitamins		
vitamin A	night blindness	part of light-sensitive visual pigment called rhodopsin
vitamin D	rickets (in infants) softening of bones (adults)	calcium metabolism
vitamin E	sterility (rats) muscular dystrophy (rabbits)	anti-oxidant
vitamin K	haemorrhage	required for blood-clotting in synthesis of prothrombin
Vitamin B group (water-soluble)		
thiamine (vitamin B_1 or aneurin)	beri-beri, neuritis of peripheral nerves	part of coenzyme for decarboxylation of pyruvic acid
riboflavin (B_2)	growth retardation (in rats) red tongue	part of coenzymes for cell oxidations (e.g. FAD)
nicotinic acid or nicotinamid (niamid)	pellegra (dermatitis, diarrhoea, madness)	part of coenzymes for cell oxidations (e.g. NAD)
pyridoxine (B_6)	convulsions (in children) dermatitis	part of coenzyme for trans-amination and decarboxylation of amino acids
pantothenic acid	no specific disease in man; growth factor for bacteria	part of coenzyme A used in fat metabolism and for pyruvic acid metabolism
biotin	deficiency sometimes produced by avidin, a protein in raw egg-white	fixation of carbon dioxide (e.g. in fat metabolism)
folic acid	anaemia	part of coenzyme for 1–C compound metabolism (e.g. in nucleic acid metabolism)
cobalamine (B_{12})	anaemia, sometimes nerve degeneration in spine	1–C metabolism, together with folic acid
vitamin C (ascorbic acid)	scurvy	involved in synthesis of collagen (connective tissue)

RECENT DEVELOPMENTS IN
THE USE OF ENZYMES

Enzymes can be up to 10 million times more effective catalysts than the inorganic ones normally used in industry. Not surprisingly, from time to time major efforts are mounted to introduce enzymes into industrial processes, with both notable successes and failures. The trouble with enzymes for industrial work is that they are rather sensitive; they are easily destroyed by high temperatures or the wrong degree of acidity. There are a few additional difficulties; in most cases it proved difficult to rescue the enzyme after the reaction had finished. This is an economic necessity; enzyme catalysts cost far more than inorganic ones, since they generally have to be extracted from the cells of plants, bacteria, or fungi. The need to produce enzymes reasonably cheaply has also limited the numbers available.

Enzymes were generally used, therefore, when a ready supply could be made available, thus lowering their price. Fortunately, such enzymes did not have to be very pure, making at least the economics of the supply side more viable. Recently some notable advances have been taking place that could mean a far greater use of enzymes in industrial processes.

These advances proceed from two directions. A number of industrial processes, notably in the oil industry, produce quantities of by-products that are difficult to sell economically. Any method that would upgrade them would be assured of a brilliant future, since the quantities involved are so large. Similar considerations also apply to any process in an industry dealing with thousand-ton quantities; plant costs are so large that the saving of even one extra process would be more than worthwhile. Thus, two interesting enzyme processes have recently been tried out by the oil companies. One of them separates waxes from crude petroleum products. Here the reason is purely economic: the separation of waxes is laborious and costly. If they could be broken down with bacterial enzymes there would be a potential saving.

The other process manufactures edible protein from oil; thus not only are there economic potentialities, but also a possible means of significant contribution to the world's inadequate food supplies. Edible protein has also been made on a pilot scale from bakery wastes and a few other by-products of industrial processes. The current problem is not so much the techniques; these are understood and could no doubt be further developed. But research in this area is conducted by large industrial companies that, not unreasonably, require at least an eventual return on the capital invested. It is said that the protein made this way would be most suitable for the less developed countries of this world where malnutrition is endemic, but at least at the moment the people in those countries cannot be persuaded to eat artificial protein. So the circle becomes truly vicious and the quantities of protein made by the costly bacteria are shunted into animal-food blending.

The other route to greater enzyme utilization is proceeding through new means of making enzymes tougher in an industrial environment. Recent discoveries indicate that enzymes could be bound to a support, such as a synthetic polymer or a cellulose. Although such a supported enzyme may lose a certain amount of activity, it still remains powerful. It also becomes far less susceptible to inactivation by temperature, acidity, or, in some cases, oxidation. It can also be recovered at the end of the process by some simple means such as centrifugation, and used again. These characteristics suggest that supported enzymes could enter some fields of inorganic catalysts and could offer viable competition, even if their initial costs were higher. One can, therefore, envisage, for example, a series of chemical transformations that would proceed by the solution being simply pumped through a series of columns or towers filled with different enzymes on their support, a specific reaction taking place in each column. In each case, the enzyme would also be recoverable.

Very recently it was reported that a much more effective method of obtaining insoluble forms of enzyme for industrial use consists in adsorbing the enzyme to silica particles and 'stapling' the enzyme molecules adsorbed on the particles together by means of a chemical reagent called glutaraldehyde. This reagent, incidentally, has been used for some years by microscopists to preserve structures, including enzymes, in tissues being prepared for examination.

TRADITIONAL USES OF ENZYMES

Enzymes are already used to an appreciably large extent, although mainly in traditional roles where at least the start of their application was rooted in craft knowledge. Thus, amylases, obtained from bacteria or fungi, are used in baking, brewing, textiles, and paper. Amylases that break down starch into simpler sugars occur in the saliva and in the pancreas. In baking, amylases break down the starch to glucose. Glucose is then further broken down to carbon dioxide by the yeast enzymes. Similarly, in brewing, amylases break down the added cereals. In the textile industry, the size used on newly made cloth contains starch, broken down with amylases. In the paper industry, on the contrary, the size needs starch broken down to form low-viscosity products.

The wonder of the chocolate with a liquid centre is also explicable. In the beginning, the centre is fairly hard. It contains the sugar sucrose plus the enzyme sucrase. After coating, the enzyme breaks down the sucrose; the breakdown products plus the other ingredients dissolve in the small amount of water present. When we eat the chocolate, we also eat the enzyme, but because it is present in such small quantities, it does not matter.

Also in the food industry, cellulose and pectinases are used to clarify fruit juices by breaking down the naturally occurring pectin, a gel-forming material. The enzymes are mainly derived from the stomachs of cattle, products of the resident bacteria that break down cellulose.

A range of enzymes are involved in the breakdown of protein. Papain, derived from the papaya plant, is used to make beer that will not go cloudy on chilling. The protein present in beer forms complexes with tannin as the temperature cools. Papain destroys the proteins and so prevents cloudiness. Another use for papain is to tenderize meat. It is also the ingredient in some tooth-paste that dissolves protein adhering to the teeth.

Ficin from figs dissolves the gelatin from scrap film and so helps silver recovery. Pepsin is used for cereal production, trypsin to prepare pre-digested baby foods, rennin to make curd. Rennin, known under its commercial name, rennet, converts milk casein to paracasein, which appears as the calcium salt, curd. Rennin is present in the gastric juices of young animals, which explains why baby vomitus always contains curdled milk.

Other minor uses of enzymes include the manufacture of cleaning powders, the loosening of hair from hides, the removal of glucose or oxygen from food preparations, and the manufacture of foam rubber; in the latter case, catalase generates oxygen, the blowing agent, from peroxide.

CHAPTER 5

→

metabolism — the systems in action

Man is a feeble creature, wrote Pascal, and he went on to enumerate all the various environmental influences that can hurt him. Even in Pascal's day, this list was quite long and since then we have managed to add a few new ones of our own. Feeble we may be, together with our fellow living creatures, yet through our evolutionary history we have devised methods of survival and action. Through biochemistry, we have the means of converting the often indifferent offerings of our environment not only to the components necessary to keep the organism alive, but also to sources of energy to give it the possibility of action.

All is movement, movement is energy, and our biological quest for survival turns into a series of attempts to acquire and harness the energy necessary for existence. This search for energy, this constant movement, the tidal waves of material passing through the organism and being manipulated, converted, turned upside down, torn apart and pieced together again form the dynamics of living. It is also what metabolism is about: activity and a vast array of equilibria all working away to allow life and growth.

All living things consist of immense cell systems; cells grow, function, multiply, die, and have to be replaced. All these events need not only the raw materials but also the driving force that ensures their incorporation into an existing or growing structure, and the removal or further utilization of waste products. We need food and we need energy, and the source of both of these is our diet.

Some living things are fortunate in having relatively simple metabolisms. Yeast cells will thrive on glucose; they have the ability to transform this sugar into all the various raw materials they need. During this transformation they can also obtain the necessary energy for all the processes they have to accomplish. In the case of most plants, energy is obtained separately from nutrition: energy directly from light, nutrition from chemically simple ingredients. So that although plants have a fiendishly complex apparatus for marrying energy with raw materials, the equation:

raw materials + energy = life and growth

can be seen in stark outline.

But when we consider animals such as ourselves we are faced with the problem that our food—proteins, fats, carbohydrates—does not contain the raw materials our cells can use in an unchanged form. We have to cannibalize its components.

We therefore break down the large molecules of our nutrients into small units, pick out the ones of use to us, reassemble them, and remove all those that are useless and potentially or actually dangerous. At the same time the energy required for growth and maintenance is obtained from these very processes. Naturally, all such systems require both chemically and physically suitable environments—chemically, because neither over- nor under-production of the required materials is conducive to a well-functioning whole; physically because at every stage there are the same problems as in any complex chemical plant: material transfer, mixing, temperature and pressure control, the supply or insulation from electricity, all have to be either at optimum values or reasonably near them. Biochemistry deals mainly with the chemical aspect of these problems, biophysics with the physical. In this chapter, we shall be concentrating mainly on biochemistry, assuming the physics of the processes to be at least sufficiently good for the job requirements to be met.

METABOLIC PATHS AND CYCLES

Fortunately for those who want to learn about the body's metabolism, it is possible to tag or mark the molecules of a foodstuff and follow what happens to them. If we do this with glucose, using a radio-active isotope of carbon, carbon-14, we shall find after a time that carbon-14 turns up in the most unexpected places, in amino acids, nucleic acids, proteins—materials that bear not the slightest chemical resemblance to glucose. We can reason that the glucose molecule has been split into its components, and these in turn have been redistributed and connected up to other molecules. This idea bears some resemblance to monetary transactions; large units are changed into small ones which can, in turn, be assembled into new groupings to give different further denominations.

This breakdown, redistribution, and reassembly in different conjunctions is the essence of biochemistry. Superficially, we can only see the gross inputs and outputs of the system. Only when we look into these processes in more detail do we see that the progress through the system is not all smooth, but proceeds along a series of interconnected pathways. These we can represent as metabolic paths and obtain a picture rather like an Underground map—of a city exceptionally well served by public transport. It is important to realize that these metabolic paths are no more than what the imagination can see; they tell us nothing about where the reactions take place, or indeed if they take place at all. But, given such a map, we can start probing it at various points and if we find what the map tells us should be there, we may begin to connect up theory and reality.

Each portion of a metabolic map illustrates a reaction and its consequences. We may say, for example, that substance A will turn into B, which in turn will produce C. All these reactions, as is the case for all metabolic processes, would be enzyme-controlled. Our map would indicate a step-by-step progression in which each product is dependent on the existence of its predecessor.

We can elaborate such a map to a considerable extent. We can consider the point at which the path branches, and a given chemical can become one of two products, under the influence of different enzymes. The daughter products continue their progress along their respective branches.

When pathways become cycles, matters become rather interesting. Whereas with straight paths once the end is reached the process stops, with cyclicals pathways the end product also happens to be the starting material. Given sufficient energy the cycle may continue indefinitely if we can disregard side losses. But if side losses do occur, material may enter the cycle at virtually any point—as it can also leave—and this usually takes place through coupling either to a straight pathway, or to any other cycle.

BREAKDOWNS IN METABOLISM

Writing down the possible metabolic pathways makes clear one fundamental idea: every step is dependent on every preceding step. If something interferes with the sequence of reactions, no steps will take place after the blockage. Two consequences follow: all the products from subsequent steps will be missing, and conversely, all the materials which in the normal course of events would have served as precursors of the post-blockage materials will suddenly appear in greater quantities than they would normally. This second effect is especially important either if the precursors would have enjoyed only a very transient existence, or if they would only have been produced in quantities too small to determine.

But how can one cause a blockage along the path? The simple solution of this problem depends on the assumption, which turns out to be correct in most cases, that specific enzymes determine the reactions taking place at each step. Therefore, if the enzyme is somehow spoiled, the step will not take place. In the main, there are two methods of knocking out enzymes which are sufficiently specific for this type of study. First, it is possible to induce, or wait for, genetic mutations which have the effect of changing the enzyme complement. Secondly, it is possible to poison the enzyme chemically. Both these techniques turn up in studies connected with the cell nucleus, although there we are interested in the exact correlation between genetic material and enzyme production rather than in the effect of the enzyme itself.

If the enzyme in a particular reaction in the metabolic path has

been made ineffective, we can observe the accumulation of precursor materials by using isotopic marking techniques. This still
leaves in doubt the actual step where the blockage has occurred,
but we can get over this local difficulty through a neat reversal of
the blockage techniques. Let us assume that we have blocked the
path from A to F at point D, and therefore stopped the production
of materials E and F. We can now artificially induce the production of these if we feed D into the system, thus replacing from the
outside a material which would have been produced automatically
as part of the normal metabolic pathway.

Although formally all the metabolic steps are reversible and
could theoretically go in both directions, in fact this does not
always happen. The metabolic pathways leading to the breakdown of material, catabolism, usually proceed through different
routes than those concerned with the eventual build-up of necessary materials, anabolism. The two therefore have to be considered
separately, despite their similarities.

THE ENERGETICS OF A
METABOLIC PROCESS

All biochemistry is, of course, only a somewhat specialized and
extremely complex branch of chemistry dealing with a multitude
of giant molecules. But if size and complexity make our task more
difficult, they should not make it confusing because, after all, a
molecule is a molecule and it is unreasonable to invest it with
magical properties only because we first came across it in somebody's liver rather than in a test tube. So we can regard ourselves
as being on reasonably firm ground when we suggest that metabolic paths are no more than a series of chemical reactions,
obeying the same rules as their rather junior partners on the
benches of chemists who specialize in simpler aspects of their
subject. In order to gain a better overall view of what goes on in
all these metabolic processes, we should perhaps recapitulate and
slightly expand some of the factors influencing chemical reaction
in general and see how these considerations fit in with what we
know about metabolism. It turns out, incidentally, that the two fit

superbly, and where they do not appear to do so, it is due to our ignorance rather than some mystical suspension of the laws of chemistry.

We should first consider the major approaches to all chemistry— that of kinetics and of thermodynamics, and the clearest way is to look again at the generalized reaction path. In its simplest case two substances react to give a third, usually called the reaction product. Since we have no particular chemicals in mind, let us call the reactants A and B and their product AB. Now both reactants and products can be considered to possess energy in rather the same way that a car does; if it is stationary the energy is potential, if it goes along it is kinetic, and if it hits a wall there is also heat, impact energy, and so forth. So one can assume that, as our chemicals react, we can in theory follow the growth and dissipation of energy, which would look rather as shown in Figure 22. There are really only two major aspects of this hump-backed bridge that are of immediate interest. First, there is the hump in the middle; a molecule has to obtain sufficient energy to get up to the top before anything can happen. If it cannot obtain this energy—which would be provided by heat or, once the re-action gets going, by the release of energy in the total reaction— nothing can happen. A and B are kinetically stable and that is the end of the matter.

But we should also notice the difference of levels at the start and at the finish. If the finishing line is lower down, the product is more thermodynamically stable than the reactants and should it get the chance to form, it will not decompose back to where it came from. But if it is at a high level, it will go back to the reactants, although it may take a very long time over it.

We can rephrase these statements and say that thermodynamics tells us which materials are more stable than others. Kinetics tells us how fast reactions will go, that is, how fast A and B would disappear, but says nothing about the stability of either the reactants or the products, that is, how much AB goes back to A and B. In reality, there are no infinitely stable reactants or products, only degrees of stability. Thermodynamics will therefore be able to give us some idea of the ratio of product to reactants in a reaction such as the ones we are examining, AB/A+B, under certain conditions, but will not tell us anything about how long such an equilibrium might take to establish itself. This dichotomy between kinetics and thermodynamics also leads us into two extreme cases. In one, the system is kinetically stable but thermodynamically unstable; the products are less stable than the reactants and would

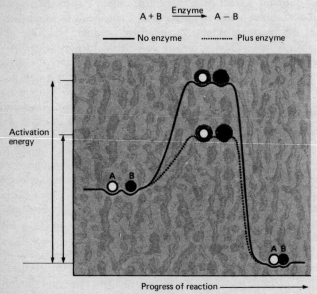

$$A + B \xrightarrow{\text{Enzyme}} A - B$$

———— No enzyme   ············· Plus enzyme

Activation energy

Progress of reaction ⟶

Figure 22   Energy Humps
Enzymes, like other catalysts, lower the activation energy and thus facilitate chemical reactions. Without the enzyme, A + B → AB might take decades to occur at body temperature.

tend to decompose, but the energy hump is too high, so nothing happens. A good example of this condition is diamond, which is thermodynamically unstable and does indeed decompose to its constituent carbon, but takes a few million years over it. The second case is that of a system which is thermodynamically stable but kinetically unstable; here the product lies at a lower level of energy than the reactants, but because the energy hump is low, there is a constant formation and decomposition of the product.

When we talk about the energetics of a chemical reaction, we mean three different things; the energy necessary to pump into the reactants so that they can reach the barrier of the energy hump, the energy released in descending from the hump to the ultimate energy level of the product, and the difference between the starting and finishing levels, which gives a measure of the total energy lost or gained during the reaction, bearing in mind that the energy level of the product may be higher than that of the starting level.

The reactions we are interested in do not, of course, happen in a vacuum. Their environment has a very important influence on the course and rate of a reaction and therefore on the energy taken in or given out. It is completely meaningless to talk about the energy required or released by a reaction without specifying its environment, and it is even more useless to talk about the energy of a molecule or parts of it, since clearly energy only has any meaning in relation to some dynamic system that includes the environment.

There is a further notion that comes into our consideration of reactions, that of order. This has been given the thermodynamic name of entropy. The laws of entropy state in effect that systems left to themselves tend to get more disorderly, and that energy is necessary to restore or increase orderliness—a common enough household experience. Changes in entropy can be measured in relatively simple chemical systems, but as yet have not been carried out in most biochemical systems because of their complexity. A result of this lack of information is that when we say that sundry biochemical systems require certain amounts of energy, we may mean that they need this in order for the reaction to proceed, for the materials to be able to climb up on the energy hump. But we may also mean that at least some of this energy is needed to restore or increase the order within the system. At this stage of our knowledge we simply cannot separate these requirements and we must assume that in most cases a dangerous degree of entropy is not threatened.

It is very possible that once accurate measurements of either change become possible some of our suppositions will look very foolish indeed.

Metabolic pathways rely on taking apart thermodynamically unstable but kinetically quite stable materials. Their components, which are thermodynamically stable but kinetically unstable, are built up into another set of large kinetically stable and thermodynamically unstable structures. The energy released during the first set of catabolic steps is used in the second stage of anabolic reactions. The simplest example is that of glucose, used as a nutrient in all organisms except plants and certain bacteria, which is oxidized to water and carbon dioxide with the release of vast quantities of energy. In the test tube, this process can be accomplished in one step, but in the living organism such a sudden release of energy would be tantamount to exploding a bomb, with very similar results. The problem is similar to one that faced the scientists who first wished to convert a nuclear explosion into peaceful uses. The energy is there, but it has to be extracted slowly enough for the surroundings to be able to deal with it and to utilize it for constructive purposes. In metabolic systems this aim is achieved by chopping up the total reaction into a great number of steps, each of which releases some energy that can be absorbed by the environment.

Similar considerations also apply to the anabolic reactions taking place in green plants. Here the raw materials are—essentially and rather inaccurately—carbon dioxide and water. These two are built up, with the help of the energy of light, into glucose units subsequently polymerized into starch and used as a store of energy for subsequent reactions needed by the plant. As is well known, if only through tooth-paste advertisements, the chemical agent through which the energy of light is trapped is chlorophyll, contained in the plant's chloroplasts.

ENERGY TRANSFER IN METABOLIC PROCESSES

Glucose synthesis in green plants also involves ATP, and since it occupies a central position both in metabolic processes and biochemical folk mythology one should make its more thorough acquaintance. To give it its full name, ATP is adenosine triphosphate, consisting of a base, adenine, a sugar, ribose, and three phosphate groups.

In green plants, synthesis of glucose from carbon dioxide and water is a complex process called photosynthesis. This initially involves the trapping of the energy of sunlight in the green pigment of plants called chlorophyll. The activated chlorophyll returns to its resting state; this releases energy which is used to split molecules of water into its components (OH^- and H^+). The hydrogen ions released are combined with carbon dioxide in a very complex series of reactions (called the Calvin cycle, after its discoverer) eventually leading to the synthesis of sugars. ATP is also required for sugar synthesis in plants. This is formed by the addition of a third phosphate group to adenine diphosphate (ADP), in a reaction which also absorbs energy released during the splitting of water. Thus, in photosynthesis, the energy of light is converted to chemical energy which is utilized for the production of sugar. The central role played by ATP in energy transformations in living systems is illustrated in Figure 23.

In these reactions a diphosphate molecule first becomes converted to a triphosphate: a reaction called phosphorylation. Under conditions existing in biological systems, phosphorylation always involves a decrease in energy of the total system, and one way, although an inaccurate one, of looking at this phenomenon is to say we have increased the potential energy of the ATP molecule. It also turns out that in other series of reactions, the third phosphate group comes off the triphosphate, converting it into the diphosphate (ADP). In turn, ADP may shed another phosphate, and is thereby converted to AMP (adenosine monophosphoric acid). This bond-breaking, again under specified conditions, results in the release of energy, which therefore becomes available for use by other molecules taking part in the reaction.

In a reasonably loose way, which would not be accepted by specialists in thermodynamics, one can say that the system ADP+P can be regarded as an energy carrier, or storehouse, because when reactions involving the addition or removal of a phosphate bond occur, there is a transfer of energy from or to the general system of reacting molecules. Such an arrangement sounds fairly high-powered, but of course there is, chemically speaking, nothing special about it. It happens quite happily with far simpler molecules in the proverbial test tube, di- and trinitrobenzene being obvious examples.

Even in biochemical systems, ATP is by no means the only example of such an energy-transfer phenomenon. When a muscle

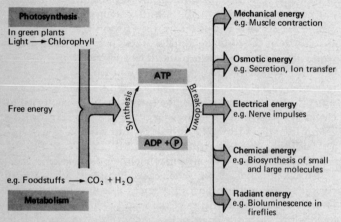

Figure 23   Energy Transformations in Living Systems are Mediated by ATP
The energy liberated during photosynthesis and the oxidation of foodstuffs in metabolism is partly dissipated as heat and partly used to convert ADP to ATP. ATP hydrolysis results in a large liberation of free energy which can be transformed into various types of energy in the appropriate transducer systems in tissues.

performs work, it needs large quantities of energy in such a short space of time that the normal energy-producing systems, based on sugar, cannot deal with it. On these occasions other phosphated molecules, for example creatine phosphate and arginine phosphate, come on the scene, where they are involved in energy-producing reactions very similar to those of ATP. It is not necessary for these 'energy carriers' to involve the breakage of a phosphate bond; there are some where a bond connecting another atom to the rest of the molecule is made or broken.

Basically, all that happens is that the formation and decomposition of particular types of molecules are attended by changes of energy of the whole reaction system, and therefore the other participants of the system suffer or gain from an energy point of view.

But why should the energetics of metabolism be of interest to anybody but the specialist? The answer is again very simple: without proper arrangements for the intake of energy from the breakdown of foods, and the subsequent availability of energy for building up molecules needed by the cells, we should not be here. And in particular, because many of these reactions do not involve oxygen, they can function in oxygen-free or anaerobic environments, which for a variety of reasons, turn out to be the most important for many biological reactions.

The availability of reactions where energy can be easily transferred also means that the total energy of food can be converted step by step in a manner most useful to the whole organism. It can similarly be used up slowly, as and when required. Thus although the total breakdown pathway must result in making available energy from food, any individual step need not be a net energy producer, a factor that introduces considerable flexibility on an evolutionary scale.

There is a further interesting sidelight on the importance of the transfers of energy in biological processes. The construction of proteins from the constituent amino acids is an energy-consuming process and one can argue that the larger the protein the more energy is needed for its construction. Very recently, work has been done to determine the age of protein molecules on the evolutionary scale and it seems that reasonable guesses can be made on the seniority of some proteins in relation to others. It turns out that older proteins are in general bigger than the newer ones, and it can therefore be argued that there is some evolutionary gain in having smaller proteins which require less energy for their formation. Specialization may not only be better for functional needs, but may also represent an evolutionary advantage in requiring less energy.

BROAD OBSERVATIONS OF METABOLISM

Somewhere along the line theories have to be substantiated and experimental data gathered for the construction of better ones. Metabolism has to be studied as it really happens. A number of techniques are now available, all possessing technical difficulties of one type or another but all having evolved to the stage where they can produce significant results. In general, the larger the part of the animal one is looking at, the more rough and ready will be the results, although in some cases, with humans having a built-in objection to being cut up or pulverized, one must usually look at the whole animal. One can therefore construct a hierarchy of size and relate it to the preciseness of results obtained.

Our first order is the whole, uncut organism. This may be as small as a yeast or as large and complex as a man, but in either case all that can be done is to draw up a balance sheet between what goes in and what comes out. In less crude terms, one can note the weight of all food ingested by the organism and observe the appearance of the final metabolites. Such an experiment cannot, of course, tell us anything about how the metabolic processes take place, but it can give some indication of how long they took under what conditions.

One can, for instance, find out if the metabolism of a certain food does take place in all species, or only in some, or only in some sub-groups or only in some individuals. The differences between such groups can then be correlated with some other characteristic. For instance, cellulose passes through the human gut unchanged, because we do not have the necessary enzymes for its conversion. One can also see how the absence of certain organs, or their deficient functioning, affects the total metabolism.

One can compare the total metabolic function of the average population with results obtained from those who, for instance, suffer from vitamin deficiency through malnutrition. Or, during pregnancy, when extra strain is placed on the mother's metabolic apparatus, hidden deficiencies may appear which would normally be masked. For example, a proportion of pregnant women suffer from lack of iron, which without the need to supply the growing foetus would not be detected. In the same way, it sometimes happens that a mild dietary deficiency can be detected because only trace amounts of metabolic intermediates would be excreted, or only very small quantities of end products derived from that particular nutrient.

In such cases it is possible to give a non-toxic precursor of the metabolite to be excreted and see how the metabolism copes with it. This is called the load, or tolerance test, and is used, for example, to test for genetic biochemical defects when these are present in such a mild form that they could not otherwise be detected. In the case of phenylketonuria, some people have sufficient enzymes to metabolize phenylanaline but only in moderate quantities; we recall that in this inherited disease a defective gene produces a defective enzyme necessary in phenlalanine metabolism. Thus if in such cases the phenylalanine load is suddenly increased, the enzyme cannot cope and abnormal metabolites will be collected in the urine, in the same way as in a full-blown phenylketonuric. The appearance of unusual metabolites in the excreta, whether under experimentally loaded conditions such as these or in the normal run of events, is usually a sign that something is going wrong in the metabolism. One can search for the missing factor. One can also determine how long a metabolite or its products stay in the organism by noting its presence in blood and excreta among other methods.

A great deal of clinical work has been and is going on along these lines; special diets ensure a good knowledge of the quality and quantity of the input, and the analysis of excreta shows the output. In complex animals such as man, these studies are often allied to medicine and pathology and can provide information on genetic defects, at least in cases when there are some overt signs of them. In simpler organisms, studies of the total metabolism can give a far more accurate picture; experiments on yeast have provided the information on the transformation of glucose into carbon dioxide and alcohol, although even in this case the intermediate steps had to await the availability of the so-called cell-free systems.

Once the total, integrated metabolism is known, the next step is to obtain some indication of where it takes place. This is accomplished by removing parts of the organism and observing what happens, or by removing blood vessels leading in and out of an organ. The removal of the liver helped to clarify its role as a major chemical factory of the body; when the liver does not function, fats are incompletely digested and even fatty acids derived from food are only partially metabolized, leading to acetone-smelling breath.

Through removal of organs, disease, or some other way of placing an organ out of circulation, one can therefore obtain some idea of the metabolic processes taking place in it. The next step is to isolate the organ but keep it alive so that one can have a better check on its metabolite balance. It is possible to feed material directly into the blood vessels supplying the organ and analyze the composition of the output. A further step is to do away with natural blood entirely, and use instead an artificial medium the composition of which is known and can be altered at will. Experiments such as these can not only reveal more about the metabolic pattern of the organ, but can also give an indication of its minimum needs, and the relationships between nutrients and organ performance. It is often found that the metabolic function of the organ is dependent on some precursor made in another part of the body, and interest is therefore switched to a preceding stage. In the same way general environmental influences can also be studied: for instance the effect of hormones on organ function.

Early experiments on the brains of cats used an artificial blood while otherwise the brain was still connected to the rest of the body. It was found that for the brain to function it was necessary to supply it not only with glucose, but also with certain other chemicals derived from the metabolism of the liver. These materials turned out to be small molecules which enable the brain cells to extract glucose from the blood.

Even when an organ is isolated, it still represents a very large mass of material. For a more intimate look it is usual to take a slice and keep it in a medium which has, as nearly as possible, the same composition as the organ's original blood supply. If the slice is thin enough, nutrients can diffuse among the cells, and metabolites diffuse out, but at the same time control over the composition of the medium and other factors can be very tight. As judged by microscopy, a sufficient number of unharmed cells survive in such a slice so that the results obtained can be directly related to what happens in the organ as a whole. Experiments like these showed the importance of the metabolism of glucose in the provision of energy for other metabolic processes taking place in the organism. The breakdown of amino acids and the body's way of dealing with nitrogen-containing wastes have also mostly been learned through slice experiments.

But there are many cases when even a very thin slice represents too much material to cope with, and in such cases even smaller parts of the original material can be obtained by breaking up the cell membranes and separating out the material they contain. Disintegrated tissue can be obtained by a variety of techniques: by grinding with sand in a mortar, by using mechanical mortars, or by an apparatus called a blender, a rather sophisticated version of the domestic mincing machine. When the cell structure is disrupted a number of particles still remain intact so that their modes of action can be studied. In a cell soup that has gone through the blender, there will be bits of the outer cell membrane, intact mitochondria, nuclei, fragments of the endoplasmic reticula, ribosomes, and so forth. Just as important, the juice will also contain the metabolites present in the cell, enzymes, and proteins, so that the sub-cellular parts, if they survived, can carry on with their usual tasks.

Once a soup free of cell membrane is obtained, it is also possible to separate out the various organelles contained in the cell, for example, the mitochondria. It is then at least theoretically possible to study the precise chemical reactions going on in a very small location, and although these reactions will no longer be under the general control of the cell, as they would be in the living organism, one is frequently able to observe possible reactions and assume that they also occur in a pristine, undisturbed state. Differences of

observed reactions between different states of organization of
tissues, for example between a cell-free system and a mitochondrial
preparation, also give important clues to materials produced at
one stage which may be necessary precursors of a following stage.

The effectiveness of all these techniques depends to a large extent
on the implied assumption that by processing tissue through rough
mechanical handling, nothing essential is changed, at least as far
as the chemistry is concerned, so that one is still dealing basically
with the starting material, rather than with an artifact. This
assumption is frequently questioned, and obviously for total per-
suasion it is important to be able to prove in a living system the
mechanisms deduced. So the task of the biochemical experimenter
is hard; from clues given by increasingly minute amounts of
material he first has to devise the scheme of a reaction which
should theoretically be valid for the organism as a whole. It also
helps if such a scheme can be tried out and seen to be working in
a completely artificial system, a so-called cell-free in vitro experiment, mainly to see if the chemistry alone is up to its selected task.
Finally, the scheme as a whole should also be capable of giving
rational explanations for phenomena observed in the living
organism. No wonder that of the many thousands of chemical
reactions and metabolic paths only a relatively few satisfy all these
criteria, and it is unlikely that future biochemists will find that all
knowledge is at hand and nothing remains to be done.

THE KREBS CYCLE

When we come to look at the finer details of metabolic pathways
we immediately hit the barrier between those who know a subject
because they are paid to do so or because they have a consuming
curiosity about it, and those who would not mind some general
outline, but, understandably enough, do not see why they should
spoil their digestion with bucketsful of meaningless jargon.
Unfortunately biochemists who find metabolic pathways interesting draw vast maps of them at the drop of a coenzyme, and the
more interested they are, the more baroque and confusing the
pathways become. It is really unfair that maps of paths are indeed

the only way to obtain some idea what is going on—in the same way that relief maps stand in for a countryside we cannot possibly visit—unless one adopts the science-fiction device of a little man who can mingle with molecules and still survive to tell his tale.

Although we shall be showing some metabolic maps, it is best to realize that details of maps are of little use to anyone but the specialist, yet general outlines are perfectly understandable even without remembering the name of every metabolite and enzyme that takes part. The beauty of the ground-plan of metabolism is that it is so simple. The main part of all metabolic processes deals with carbon atoms and bits of other groups attached to them. Irrespective where they start all end up with a compound called acetyl-CoA. For the purpose of this argument it does not greatly matter what this material is, but it will help to note that the end that bites contains two carbon atoms. Whatever food we start off with, this molecule is always the end result of a substantial portion of the breakdown processes. The starting material may be a polysaccharide such as starch, proteins from meat, or fats, and obviously all these materials contain other chemical groups which have their own ways of being broken down. But by far the largest part of metabolism deals with materials containing only carbon, oxygen, hydrogen, and nitrogen, and it is essentially these processes that provide the powerhouse to keep the whole organism going. Later on in this chapter we shall consider how the various foodstuffs are broken down into their respective moities, but if we only focus on this central portion of metabolism we are faced with acetyl-CoA.

Coenzyme A is a carrier of two-carbon bits in a somewhat similar fashion that, as we shall see, some RNAs are carriers of amino acids. The two-carbon bits coming off the previous catabolic processes can be routed into various other processes. They can be started off on their way to become sugars, or they may become parts of the build-up of fatty acids. They may be used in acetylation, a process used to inactivate toxic drugs, and also to build up acetylcholine, the chemical in the transmission of impulses through nerves.

But the most important function of the coenzyme A carrier is that it funnels the two-carbon bits into the citric acid or Krebs' cycle. This giant system of cyclic reactions, elucidated by Sir Hans Krebs and his collaborators, drives the whole metabolic system to produce reactants for energy-producing reactions. The Krebs cycle begins with the condensation of a two-carbon bit, carried by coenzyme A, with an acid containing four carbon atoms called oxaloacetic acid. The product is six-carbon citric acid. Citric acid then goes through four oxidizing stages, but instead of oxygen being added on, hydrogen is taken off. These reactions, as always in metabolic processes, are enzyme-controlled; their technical name is dehydrogenation. At two steps carbon dioxide also comes off. Altogether the cycle contains seven steps. At the last step we end up with a molecule containing four carbon atoms, and this is of course where we first came in, because oxaloacetic acid, the acid containing four carbon atoms, is precisely the bit that is attached to the two-carbon fragments entering the cycle. Provided that sufficient enzymes, intermediate metabolites, and energy are available to keep the cycle turning, this one central process supplies all the raw materials necessary for the last great act of the metabolic drama, the burning of the breakdown products to obtain the necessary energy for the rest of the organism's metabolism.

It is permissible to simplify and to consider that the Krebs cycle produces only hydrogen atoms and carbon dioxide. It obviously does much more, but it is sufficient explanation for the subsequent processes called cell respiration if we consider only these two. Our account has to deal with the fate of hydrogen and carbon dioxide.

The actual physical reality of these processes takes place in the mitochondria of the cells. We have observed that these organelles are well suited to materials handling; they have a very large surface area where reactions can conveniently take place and, being mobile organelles in the cell, they have direct access to metabolites entering the cell or to products of reactions occurring in the cell, for instance, pyruvic acid derived from glucose. The fate of carbon dioxide is simple; most of it is carried out to the lungs by the red blood cells and is exchanged for oxygen from the atmosphere. In other words, we have to account for the hydrogen produced in the dehydrogenation reactions plus the oxygen that has appeared as a result of respiration.

It would be possible to oxidize hydrogen to water in one step as indeed is the case in many reactions outside biological systems, but the energy jump would be so high that the surrounding system would be unable to utilize fully the released energy. So instead we have a great number of steps, in each of which an oxidation-reduction reaction takes place, with some releases of energy and some degradation of the raw material. Indeed, instead of oxidizing hydrogen molecules in their entirety, these processes take them apart into electrons and protons, and both of these proceed along their appropriate paths.

The carrier bringing the hydrogen atoms from the Krebs' cycle is the reduced, or hydrogenated variety of a coenzyme called NAD. The reduced coenzyme transfers a hydrogen atom to another carrier, a species of protein called flavoprotein. The flavoprotein has yet another transport colleague, called ubiquinone. But after the ubiquinone stage the load is spread: the electron and the proton of the hydrogen atom part company. The electron is shunted across three stages provided by another set of pigmented proteins, called cytochromes. At the terminus they are joined by protons coming straight off the ubiquinone and by oxygen that is ultimately derived from the atmosphere. All the components being present, water then forms.

But the respiratory chain has another fundamental consequence: the creation of ATP that, as we shall see, turns up in a variety of reactions that are net users of energy. At each stage of the respiratory chain—NADH, flavoproteins, ubiquinones, cytochromes—energy is liberated. Concurrently, the energy-requiring synthesis of ATP from ADP phosphate takes place. This reaction is called phosphorylation. The stages of respiration involve the oxidation, or dehydrogenation, of one species and the reduction or hydrogenation of another. Thus the total system is described as oxidative phosphorylation, and as we shall see (cf. Chapter 7), the coupling between them provides a control mechanism for cell respiration. The exact nature of the coupling mechanism is still unknown, although we shall be discussing two current theories explaining the way it functions.

Both the cytochromes and haemoglobin of the blood use iron as a carrying agent. Recent work has tended to show that cytochromes are much older proteins on the evolutionary scale than haemoglobin. Indeed, if one compares cytochrome molecules at various stages of the evolutionary ladder, we can follow the gradual simplification of this molecule and the paring down of presumably disposable parts: the number of amino acids falls, for

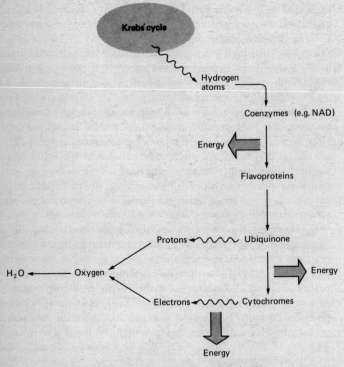

Figure 24   Hydrogen Transfer in Respiration
Energy released is absorbed in the synthesis of ATP from adenosine diphosphate and phosphate.

example, from over 120 in the wheat germ to just over 100 in man. So it is interesting to realize that one very old protein and one relatively new one both use the metal iron, presumably because of its particular suitability in biological environments for the rapid making and breaking of certain types of bonds.

Overall, there are three main points that emerge from our rapid review of the chief features of the total catabolic mechanism. The most important is that although there is a clear energy relationship between starting materials and reaction products (and indeed in many cases the latter could be obtained from the former in one step), in biological systems a related series of reactions takes place. These allow for the gradual absorption of energy released into the environment and therefore the creation of molecules which, when taking part in further reactions, can create a new environment where the release of energy will occur.

For such a series of reactions to take place, it often happens that the main entity being transformed becomes attached to some other molecule or parts of it; acetic acid, for example, becomes involved with a coenzyme before entering the Krebs' cycle. There is a great deal of such co-operatives or temporary mergers taking place in all types of chemical reactions occurring in biological systems, since the complex entities thus formed have a better possibility of undergoing particular reactions. These associations dissolve immediately as soon as the particular reaction has taken place. The secondary, or unreacted part then usually enters into a new set of reactions; one can think of them as special types of internal catalysts. Another way of saying the same thing is to describe them as activators, allowing the molecule being activated to take part in a specific reaction. Both these descriptions are essentially inaccurate and from a precise point of view probably misleading. But they do give an idea of what is going on and the description 'activated molecule' is quite generally used in biochemical literature.

Finally, it is interesting to note that the transfer of material, whether molecules, bits of them, or electrons, takes place along the links of a chain. The chain consists essentially of pairs of molecules handing to each other one particular atom: either a hydrogen or a phosphorus. There are, of course, a great number of detailed differences between these processes, but in all cases it is the handing over of a particular atom that allows the metabolic product to proceed along the chain, and conversely the appropriate energy transfers to be made between a reaction system and its environment.

SUGAR METABOLISM

It is disingenuous to try to apportion the importance of the various metabolic cycles occurring in an organism, but historically the conversion of sugars was elucidated first, and as it turned out, it is of very great importance in supplying the body with energy. We shall therefore examine what makes the ingested sugar turn into acetyl-coenzyme, the point at which the Krebs' cycle takes over, and in general what is the significance of these particular mechanisms.

It can be shown by tagged carbon experiments that sooner or later ingested sugar finds its way into practically all parts of the body. It has also been shown that some tissues, such as the brain, take their energy supply directly from sugar served up to them in the blood supply. Sugar metabolism therefore occupies a central position in the working of the whole organism.

But the sugar metabolism story started long before carbon tagging experiments: in fact it was Pasteur who studied yeast fermentation, although he explained it on the basis of the since discredited vital principle. Somewhat later, the Buchner brothers in Germany found that yeast juice alone has the capability of fermenting sugar, but the ability was lost if certain small molecules, which later were identified as coenzymes, were separated out. At this stage we therefore had the knowledge that yeast and similar organisms possess one or more factors, later called enzymes, capable of converting sugar to alcohol, and that these factors also need the help of other chemicals to carry out their tasks properly. Just how many of these helpers were required was becoming clearer around the turn of the century from the work of Harden and Young, who found that although cell-free yeast juices could start fermenting sugars, they soon ran out of steam if they were not supplied with phosphate. This observation laid the groundwork for a very large amount of subsequent experimentation resulting in a fair understanding of the importance of the phosphate-coupled 'activated' molecules. We should also consider, perhaps with some awe, that the main outlines of sugar usage are exactly the same all along the evolutionary ladder, and differences between our utilization of sugars and that of the most humble yeast cell are far less than the similarities—no doubt a tribute to the evolutionary excellence of these processes.

The starting material for the breakdown of sugar is glucose, either obtained directly from food through the digestive processes, or taken from storage in the tissues, where it can be stockpiled as glycogen. The complete process of converting sugar to carbon dioxide and water is called glycolysis and, if we wish to be accurate, there are quite a number of metabolic paths, although only two are of major importance. These are called, from their discoverers, the Embden-Meyerhof path and the Warburg-Lipmann-Dickens or pentose path.

The first of these, the Embden-Meyerhof pathway, does not require any free oxygen and is therefore an anaerobic process, shared by organisms in every part of the spectrum of life. The dual foundations of the pathway are a vast array of enzymes, which control each particular reaction system, and the intervention of ADP and ATP to supply or transfer phosphate residues. The starting material, glucose, is converted into phosphate derivatives which then go through a series of molecular gyrations, and are finally broken down to three-carbon units. The end result of these processes is that where we first had a glucose, we end up with two sugars, each containing three-carbon atoms and one phosphate group attachment. This species, triose phosphate, acquires an additional phosphate group in a most unusual way: the phosphate does not come from ATP, but from free inorganic phosphate.

On the scene now comes a coenzyme, NAD. We have already met it in the respiratory cycle, and here it also fulfils the same role; it is an oxidizing agent and carries off a hydrogen from the phosphorylated sugars. This opens the way to the shedding of a phosphate group, taken up by a molecule of ADP, so at least some of the ATP which became ADP in the first part of this cycle has now been regenerated. Further processing converts the 3-phospho-glyceric acid produced into pyruvic acid, accompanied by the shedding of yet another phosphate group on to a waiting ADP. The presence of ATP has a twofold significance: these processes are net creators of usable energy, although they are not very effective, and the main source of energy is further along after the Krebs' cycle. However, ATP also functions as a control on the amount of glucose being broken down, and it is interesting to note that after physical exertion, when a great deal of energy is used up, the system will go on functioning in top gear up to the point where all the ATP used is regenerated, and the pathways are again fully serviced.

Pyruvic acid produced by glyolysis is reduced by the coenzyme $NADH_2$—formed previously in the sequence. In man, pyruvic acid turns into lactic acid, in yeast to acetaldehyde and subsequently to alcohol. Such a reduction reaction assumes that the production of pyruvic acid and the reduced coenzyme are closely coupled. If the presence of air disrupts the amounts of available materials, some coenzyme is re-oxidized in the respiratory chain. Not all pyruvic acid will therefore be converted to its usual breakdown products. Instead an enzyme, pyruvic acid oxidase, comes in on the attack and converts the acid into carbon dioxide and a derivative of coenzyme A, already discussed in relation to the Krebs' cycle.

In the other sugar metabolism pathway, the pentose or direct oxidation pathway, oxygen takes at least an indirect hand. The chain starts with glucose-6-phosphate, which is twice oxidized by a coenzyme similar to NAD. The coenzyme in the Warburg pathway is NADP—NAD with an extra phosphate group—and after it has finished oxidizing the sugar it is twice reduced, becoming $NADPH_2$. The reduced species must therefore be re-oxidized, a reaction carried out in the respiratory chain, during which oxygen is consumed.

Material going through the complete Warburg pathway can re-enter the other glycolysis sequence as pyruvic acid. But the main importance of this route is that it provides the starting points of a number of derivatives for building nucleic acids and other necessary molecules. It is also important for plants, and indeed it was in plants that it was first studied.

PROTEIN METABOLISM

The beginnings of our understanding of protein metabolism follow the same general outlines as has been the case with carbohydrates. At certain times proteins came to be regarded as the most important materials in the living organism, only to be dethroned recently by nucleic acids. However, proteins, and their precursors, the amino acids, are certainly of very great importance, and a great many elegant pathways exist to see that we obtain the requisite number of amino acids necessary for their synthesis.

Protein metabolism is therefore a story of what happens to amino acids in the organism. In 1939 Schoenheimer observed that if amino acids tagged with nitrogen are fed to animals, the marked atoms reappear in the urine within twenty-four hours, mainly as urea. But not every one of them; some stay longer, and it became obvious that something fairly complicated was happening to the ingested amino acids. This observation was underlined by the fact that the tagged nitrogen atoms also turned up in tissue proteins, and some in proteins of non-dividing tissue, for instance in the muscle, bone collagen, and connective tissue. From observations of this type came about the concept of the dynamic equilibrium of amino acids in the body, so that today we believe that irrespective of whether the cells in a given tissue divide or not, their protein content is exchanging amino acids with the environment at rates specific to the tissue. For example, amino acid turnover in bone collagen is very slow, in muscle about middling, and very high in tissues such as the brain, the liver, and the pancreas. The breakdown products are amino acids that after losing ammonia join the urea cycle and are excreted in urine.

As far as the catabolism of protein molecules is concerned (whether they are already in the cells, or whether they are ingested), the organism possesses mechanisms for their breakdown into amino acids. In the cell a set of special enzymes, called cathepsins, can break down proteins. Cathepsins are found both in the lysosomes and in the rest of the cell. Proteolytic enzymes are also protein-breakers. As we have seen, there are digestive mechanisms which break down the proteins ingested in the organism's diet. In any case, the end result is amino acids whose fate we now have to follow.

As in carbohydrate metabolism, there are a number of catabolic paths for amino acids, of which we shall look at the two most interesting ones: transamination and oxidative deamination.

In transamination an NH_2 group, called an amino group, transfers from an amino acid to a keto acid. The amino acids are intermediates in the glucose breakdown sequence, whereas the keto acids appear in the Krebs' cycle. Thus transamination can fulfil a double task: in addition to its catabolic role it also provides a bridge between glucose breakdown and the Krebs' cycle. This role is determined by the reversibility of the reactions; the Krebs' cycle can produce amino acids, but amino acids can also feed back into the cycle.

Both the main amino acid breakdown procedures use coenzymes as carriers. In transamination it is pyridoxal phosphate, a derivative of vitamin B_6, in oxidative deamination it is flavin-adenine-dinucleotide or FAD.

There are also a number of other amino acid breakdown pathways, some with interesting products. For example, some amino acids can be made to lose carbon dioxide to give an amine. One of these amines, tyramine, turns up in cheese. It has a vaso-constricting effect, that is, it constricts the blood vessels. Thus cheese eaten before going to bed can result in nightmares. Too much tyramine is toxic, especially to brain tissue, and must be quickly removed from the organism. Another amine, dopamine, is a precursor of the neurohormones, while yet another, serotonin, is probably a neurotransmitter, involved in nerve conduction.

Some other decarboxylated amines are formed in the gut through bacterial action; they rejoice in names such as putrescine and cadaverine, and one can leave it that these names are sufficiently descriptive.

An interesting point in the breakdown of amino acids is the formation of ammonia in some cases. Ammonia is highly toxic and cannot be allowed to remain in the organism, even for the time needed for transport to the kidneys. Thus ammonia is quickly carried away in the urea cycle.

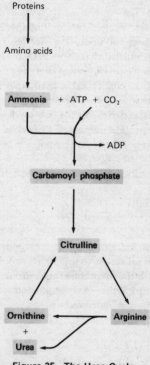

Proteins

Amino acids

Ammonia + ATP + CO$_2$

ADP

Carbamoyl phosphate

Citrulline

Ornithine ← Arginine
+
Urea

Figure 25 The Urea Cycle

In mammals, urea synthesis proceeds through a circular arrangement, which meets the necessity that sufficient reactants should always be available for the removal of toxic materials. We have again one of the enzyme-assisted reactions which include ATP among the reactants. ATP sheds a phosphate to become

ADP. The ammonia, together with carbon dioxide from the respiratory chain, becomes a molecule called carbamoyl phosphate. This goes through a number of reactions (as summarized in Figure 25) and along the line urea is produced, which can then be removed in solution. The important part of these proceedings is that ornithine is regenerated so that the cycle can start again. An interesting point to note is that the urea cycle, in the same way as the central citric acid cycle, was elucidated by Sir Hans Krebs.

The total synthesis of urea is a net energy consumer; one measurement of this fact is that three ATP molecules are required to produce one molecule of urea. Whereas urea is the end as far as nitrogen wastes are concerned in mammals, in plants and some lower animals there are further enzymes which split it into ammonia and carbon dioxide. As we have noted already, the evolutionary justification for such a mechanism resides in the availability of water to carry the toxic ammonia out of harm's way.

There are a number of odd features about our use of amino acids. For example, there are twenty-two of them necessary for the living organism, yet we are able to produce only a few of them; the rest have to be ingested. Yet again, all amino acids are optically active and, as we have seen in Chapter 4, only the L-form is used by the organism. In metabolic terms, there is an enzyme whose special task is to inactivate any D acids, since these could interfere with the metabolic mechanism. There are no complete explanations for either of these phenomena, only the shadow of a still unexplained evolutionary past.

FAT METABOLISM

As most of us are forced to realize sooner or later, the body is good at storing fats. As a matter of fact, it does this by collecting molecules of fat in the tissues in a form that chemically resembles soap—long-chain fatty acids, which in a soap would be attached to a metal, when stored in the body are combined with an alcohol. More precisely, fats are stored as glycerol esters. When they are needed, the bond between alcohol and fatty acids is broken by an enzyme that is in turn produced in response to hormone control. After this stage we are merely following the breakdown of the fatty acids until their products are funnelled into the Krebs cycle.

The first fundamental—and as it turned out to be, correct—observations on fat metabolism were carried out at the end of the century by the German physiologist Knoop, who fed dogs fatty acids which were labelled not with radioactive isotopes, not recognized for this type of work at the time, but with a phenyl group. If a phenyl group is attached to a fatty acid, the bond formed is so strong that the fatty acid part can go through a variety of chemical reactions and the phenyl will not come off. When in the end we wish to test for the presence of what had been a fatty acid to start with, we only have to make sure that we find the phenyl group. This is reasonably easy to do analytically.

Knoop fed marked fatty acids to his dogs and observed that whenever the fatty acid contained an even number of carbon atoms in the chain, phenyl aceturic acid appeared in the dogs' urine. This was clearly an indication that fatty acids were being degraded to phenylacetic acid, since it was known that phenylacetic acid is always excreted as phenylaceturic acid. Similar arguments also applied when the fatty acid fed to the dogs contained an odd number of carbon atoms in the chain. This time the end product was hippuric acid, a benzoic acid derivative containing glycine. In both cases, therefore, the fatty acids ingested must have been broken down in the animal, and Knoop postulated a sequence consisting of a minimum of three steps, involving oxidation.

Subsequent experimentation in this area consisted in verifying, refining, and simulating this experiment in cell-free systems. Its essential rightness was very amply proven. Finally, the complete sequence of events could be reproduced through identified raw materials obtained from mitochondria, in which these processes take place, a biological engineering feat associated with the name of D. E. Green in the U.S.A. and with F. Lynen in Germany.

What happens is that, rather as in the case of carbohydrate metabolism, fatty acids become 'activated' by ATP and are then transferred to coenzyme A. In technical language, the fatty acid thioester of coenzyme A is produced and in this 'activated' form goes through a series of steps in which hydrogen atoms are stripped off by their carriers, and water is added, so that in the penultimate stage of the process we do obtain the entity postulated by Knoop in his dog experiments.

Thus, the fatty acid is stripped of two-carbon fragments along the chain: these are processed into the Krebs' cycle through the usual routes. The major portion—still a fatty acid although rather smaller—continues along the cycle. At the re-entry point the diminished fatty acid, still attached to Co A, already primed, can rejoin the cycle. The whole system is called, by only a small stretch of imagination, the fatty acid spiral.

These processes also involve the ability of flavin coenzymes and NAD to take up and shed hydrogen atoms; thus fatty acid oxidation is coupled through them to the respiratory chain. Overall, the breakdown of fats generates energy which then can be used for synthesis. For purposes of comparison one looks at the amount of energy generated by the breakdown of one gramme of fat, of sugar, and of an average protein. The units in all cases can be expressed in terms of ATP molecules.

SYNTHESIS

These examples have shown something of the processes through which material, either ingested or already in the system, is broken down so that unwanted portions can be removed, energy obtained, and valuable raw materials provided for subsequent synthesis. At one time it was thought that synthesis was merely the reverse of catabolism, since, at least theoretically, all the catabolic reactions are reversible. In practice, this does not happen. Why it should be so is not quite clear, but it is certain to be connected with the energy requirements of the reaction systems involved. However, the basic principles of the synthetic reactions are the same: they also proceed through the intervention of catalyst, carriers, and activators. For a professional biochemist, the knowledge of the precise nature of these reactions is obviously of vital importance, but there is not enough significant difference between them for us to look in detail at every one of them. In order to see the general scheme of the type of metabolic pathways that syntheses often take, we can look at the processes that build up fatty acids. Let us consider a medium-size representative, a fatty acid with sixteen carbon atoms.

This process is actually rather interesting because it takes place on a large multi-enzyme complex which acts as a template or guide to the deposition of consequent portions of the acid. Although the chemistry itself is quite different, the idea has something in common with the synthesis of proteins, where another type of template, the ribosomes, control the sequence of deposition. The starting materials in fatty acid synthesis are two coenzyme complexes: acetyl coenzyme A, and another similar entity, with the difference that instead of the two-carbon acetyl portion we have the three-carbon malonyl moiety. The acetyl coenzyme transfers its acetyl group to the enzyme complex and subsequent three-carbon portions of the fatty acid are delivered to an adjacent site by malonyl coenzyme A. The acetyl coenzyme A is only used initially, and does not take part in further reactions. Once the acetyl portion is lodged on the enzyme site, a number of other reactions also start to take place, resulting in the condensation of two-carbon units with simultaneous evolution of carbon dioxide, oxidation-reduction sequences and the regeneration of the carrier molecules involved in the processes. Instead of taking place in the mitochondria, as do the breakdown processes, fatty acid synthesis occurs in the cell cytoplasm; therefore, the actual place of synthesis does not appear to need the local specificity of the mitochondria. It has also been found through work with cell-free systems that the raw materials alone will not come together on their own to form the final products. ATP, manganese ions, the vitamin biotin, and carbon dioxide are all necessary for the synthesis to go ahead. As suggested, these materials either act as carriers of the appropriate molecular portions in the synthesis, provide the reaction systems with the necessary energy, or, in the case of the metal ions, afford the active places on the enzyme complex where the necessary catalytic function is carried out.

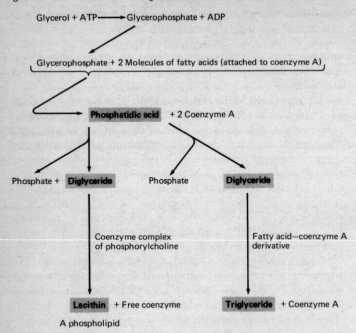

Glycerol + ATP ⟶ Glycerophosphate + ADP

Glycerophosphate + 2 Molecules of fatty acids (attached to coenzyme A)

Phosphatidic acid + 2 Coenzyme A

Phosphate + **Diglyceride**　　　Phosphate　　　**Diglyceride**

Coenzyme complex
of phosphorylcholine

Fatty acid—coenzyme A
derivative

Lecithin + Free coenzyme　　　**Triglyceride** + Coenzyme A

A phospholipid

Phospholipid synthesis　　　　**Triglyceride synthesis**

**Figure 26　The Synthesis of Phospholipids
and Triglycerides from Glycerol**

But what happens to the fatty acids once they are formed? They become the starting materials for the synthesis either of more complicated lipid molecules or are transferred into a state suitable for storage. If we take glycerol as an example, it is first phosphorylated by an ATP molecule, exactly in the same way as glucose was by ATP in glycolysis. The alcohol groups involved can then take up two fatty acid molecules and a phosphorus atom, forming what is called a phosphatidic acid. These reactions, which occur through the intervention of the acetyl coenzyme A, mainly take place in the small intestine, and in the liver, and of course in the adipose tissue, since they provide the precursors of the triglycerides, the form in which fat is stored. Triglycerides are formed simply by the substitution for the phosphorus atom in the phosphatidic acid of another fatty acid molecule obtained from a coenzyme complex.

There is also a second route from phosphatidic acid which, on going through a diglyceride and shedding a phosphate in the process, results in the production of phospholipids. Phospholipids are essential components of cell membranes and in gland tissues are also probably connected with protein synthesis.

CHAPTER 6

genetics and replication

The central problem of modern biochemistry concerns our knowledge of factors governing the birth, life, and death of cells. To many people, both within and without the sciences, this is biochemistry. When we have examined as far as our present knowledge permits the whole gamut of functions carried out by the living organism, we arrive at a number of problems concerning individual cells, irrespective of whether our methods involve an actual living organism or some simulated, in vitro system. Cell behaviour, in turn, can be regarded as being determined by a whole complex of processes proceeding in the control centres provided by the nucleic acids.

We can argue that what a cell is, is in a sense, what a cell does. Knowledge concerning nucleic acids is therefore tantamount to knowledge of cells. We shall consider the role of nucleic acids in two parts: the passing on of genetic information, and the control of protein, and especially of enzyme, synthesis. This scheme is purely a matter of convenience since these two functions are merely two manifestations of the nucleic acids' interactions with their environment. We shall perpetuate a further logical short circuit; instead of following historically the processes of discovery leading to our present knowledge, we shall start off with the present and flash back to the more important factors leading up to it. The edifice of molecular biochemistry is complicated enough even with a ground plan; without it, even an expert can take the wrong turning.

Despite the recent growth of knowledge we must beware of taking statements too much on trust. On the following pages we shall be looking at some of the current theories concerning nucleic acids that we believe to be true. It is possible, but unlikely, that in the coming years most of these statements will be proven false; more than likely many a detail will have to be modified. This constant flux of new knowledge, differences in results and interpretations, and the overall complexity of the problem make nucleic acids fascinating both to the professional scientist and to the outsider, who can only appreciate the outlines of this intellectual edifice.

The ways of understanding nucleic acids are like the routes mountaineers use to reach the summit: some are steep and short, some long and winding. All require considerable effort and perseverance. So in biochemistry there are shifts of emphasis depending on the point of view of the experimenter; molecular biochemists, geneticists, virologists all have their special interest. There are many special topics that make up the complete argument about the nature of nucleic acids, and although of necessity these have to be presented separately, it must be borne in mind all the time that one is still talking about the same substances.

We can oversimplify and differentiate broad areas where nucleic acids appear to have an important role: the inheritance of characteristics and the control of cell functions, or what materials the cell manufactures under which conditions. We should also note that most research work presenting us with our current knowledge has been carried out on very small organisms indeed, since very large numbers of them can be bred identically and also possess very short life cycles. Every time a piece of nucleic acid is synthesized in a test tube, or probably in a rather complicated glass apparatus, it is hardly a signal for portentous warnings about the imminence of artificial humanoids. We are still rather a long way from that.

HEREDITY

In the middle of the nineteenth century an Austrian monk named Mendel performed a vast number of experiments to find out how sweet peas inherit observable characteristics such as colour. The immediate effect of his work was nil. Mendel's experiments were forgotten, but by the end of the century came rediscovery and a general consensus that characteristics are inherited in small, discrete packages. These collections of packages later became identified with actual physical entities in the cell nucleus which could be stained and viewed under a microscope: the chromosomes or coloured bodies.

We can now translate these observations into our current image of inheritance; the chromosomes consist of the whole or parts of one of the key nucleic acid molecules called DNA, together with proteins and fat-type lipids. The functions of the DNA molecule are now known in fair detail. The role of the proteins is still not entirely clear and we are ignorant of why the lipids should be there. The actual inheritance package is called the gene, and we can now equate it with discrete portions along the backbone of the DNA molecule.

Cell biology describes two types of cells, reproducing in different ways. Sex cells undergo a type of division called meiosis, in which the daughter cells receive only half the genetic material contained in each parent cell. On fertilization, two sex cells, containing only one chromosome each and therefore described as monoploid, pair their chromosomes. Pairing results in a cross-over of the two original units. The new chromosomes, which now belong to the daughter cell, form from the two loops of the crossed-over pair. Thus the daughter cells contain genes from both parents; the amount of material inherited from either depends on the crossover point and where the two chromosomes had separated.

Somatic 'body' or non-sex cells contain a double set of chromosomes. During division, half of each member of a chromosome pair is segregated in opposite ends of the cell, followed by chromosome division. Thus the total number of cells is doubled, all daughter cells being identical to the parent. This process is called mitosis.

Obviously, there must be a reason for meiosis and mitosis. One can argue that meiosis is necessary for survival since the crossing over and subsequent separation of chromosomes allows the random redistribution of characteristics and therefore the emergence of the most viable individuals. It is certainly remarkable that sexual reproduction can occur at very low levels of life; even bacteria can take part in a rather primitive form of it. This is called conjugation and involves one bacterium attaching itself to another and transferring some or all of its chromosome material.

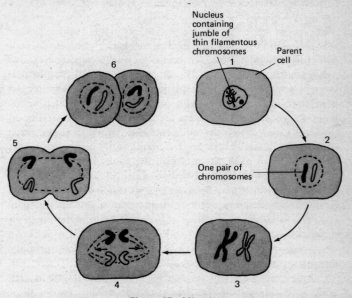

Figure 27   Mitosis
One pair of homologous chromosomes followed through mitotic cycle. Chromosomes are duplicated in (3) into two identical 'chromatids' and each half of the duplex is segregated to opposite ends of the cell. Note that the nucleus' membrane no longer exists. This is later followed by cell division. Thus, somatic cells divide by mitosis into identical daughter cells (6), each containing the same chromosome number as the parent.

Not surprisingly, the one that carries out the transfer is called the male bacterium.

In both these cases of replication, it is necessary that actual division should take place accurately. Both in the formation of a new individual—meiosis—and in the growth of the body or the regeneration of cells—mitosis—it is vital that continuity should be preserved; otherwise the species or the individual would soon die. Mistakes are exceedingly rare. We can therefore stipulate that if nucleic acids really fulfil some function in cell division and pass on genetic information, they must be constructed to accomplish this accurately.

STRUCTURE OF NUCLEIC ACIDS

There is one major feature common to nucleic acids, trees, and nylon stockings, among other things. They are all composed of large polymeric molecules. While in the usual run of chemistry we tend to come across relatively small molecules—water, acids, bases, hydrocarbons—polymers are the giants of chemistry, many thousands and sometimes millions times larger than the water molecule. Although there are enormous variations between sizes and characteristics of different polymer molecules, they all conform to a reasonably simple ground plan. All polymers consist of a backbone, which may be straight, zigzag, or curved. Attached to the backbone are a number of side groups. All nucleic acids conform to this pattern, their differences being determined by differences in composition and arrangement. The latter is described in the terms already used for proteins: primary, secondary, and tertiary structures.

The largest of the nucleic acids is DNA, or deoxyribonucleic acid. Its backbone consists of the sugar, deoxyribose, and a phosphate. The side groups, attached to the backbone are four large nitrogen-containing ring structures called purines and pyrimidines. As we shall see, the relationships between these side groups both provide structural stability for this enormous entity and also determine its action.

About 99 per cent of the cell's DNA is found in the nucleus, the remainder residing in the mitochondria. In the nucleus it is associated with a series of basic proteins called histones, other proteins that are acidic, and the fat-like lipids.

Mitochondrial DNA appears to be arranged in several loops, the so-called hypercircles. It was also found that DNA exists in the photosynthesis units of green plants, the chloroplasts, and these two observations have suggested that much extra-nuclear DNA may be self-replicating. Such a train of reasoning leads us to the very beginning of life on this planet, since it is argued that the quasi-independent DNA particles outside the cell nucleus bear a close resemblance to ancient bacteria that, in contrast to their modern offspring, were capable of synthesizing their own structural and functioning proteins. Thus it is possible that the extra-nuclear DNA in the cell may be of a more ancient lineage than the nuclear DNA and may have evolved from the ancestors of our present bacteria.

DNA is usually prepared from calf thymus gland since these cells contain little cytoplasm and there is therefore less material to process. But it has been found in virtually every type of cell, and it is therefore reasonable to assume that DNA is a basic necessity all through the living world. Exceptions are the red blood cells of several vertebrate species, including man, although an exception to the exception are the red blood cells of birds, which are nucleated.

The DNA molecule is enormous. It has been found with a molecular weight in the region of a thousand million compared with the weight of carbon 12 and in at least one bacillus, *E. coli*, it has been found to be nearly 1.0 mm long. Since the length of the whole cell of this bacterium is only about one thousandth of a millimetre, it is obvious that in at least this case the DNA molecule must be tightly coiled to fit within the available space.

There are three main varieties of smaller nucleic acids: all of the type RNA, ribonucleic acid. These are also polymers with a sugar-phosphate backbone, but whereas in DNA the sugar is deoxy-ribose, in the RNA chain ribose sugar is used. The 'bases' are identical to those of DNA, except that uracil takes the place of thymine. The three variations of RNA are found in three different places in the cell, performing different functions. About 80 per cent of the RNA complement is outside the nucleus, associated in structures called ribosomes with a number of proteins whose functions are being currently investigated. This is called ribosomal RNA or rRNA.

A less stable form of RNA, comprising less than 5 per cent

(1) Base pairs

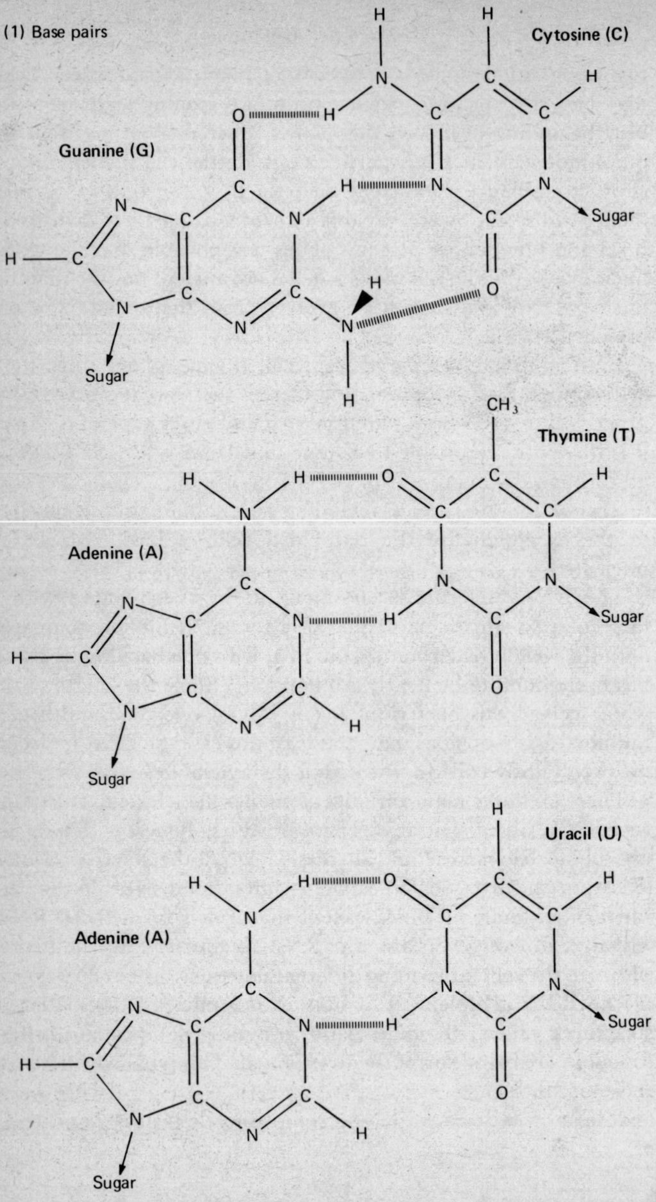

Cytosine (C)

Guanine (G)

Thymine (T)

Adenine (A)

Adenine (A)

Uracil (U)

(2) Sugars

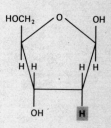

Ribose   **Deoxyribose**

(a) *Base pairs and sugars in nucleic acids.*
Nucleic acids are polymers containing four main types of base attached to a 'backbone' of sugar-phosphate residues. The acidity of nucleic acids is due to the phosphoric residues of the 'backbone'.
(1) Bases form pairs through hydrogen bonds (ıııııııı) between a 'purine' base (adenine, guanine) and a 'pyrimidine' base (cytosine, thymine). In RNA, thymine is replaced by uracil.
(2) Ribose is the sugar in ribonucleic acid (RNA); in DNA (deoxyribonucleic acid), the sugar, deoxyribose, contains one less oxygen atom.

Figure 28(a)   The Structure of Nucleic Acids

of the total cell RNA, is also found on the ribosomes. For reasons we shall be looking at further on, it is called messenger RNA, or mRNA. From 10 to 15 per cent of the RNA is found in the soluble cytoplasm and is called soluble sRNA or transfer tRNA, according to whether we wish to emphasize its provenance or its function. Finally, about 1 per cent of the total RNA resides in the mitochondria, and as in the case of RNA associated with the nucleus, is involved in protein synthesis.

Until the 1930s the nucleic acids were no more than chemical curiosities, and one can also suggest that experimental techniques were not sufficiently developed to deal with such enormous molecules in a meaningful way. However, in the 1930s and 1940s two pieces of important experimental evidence came

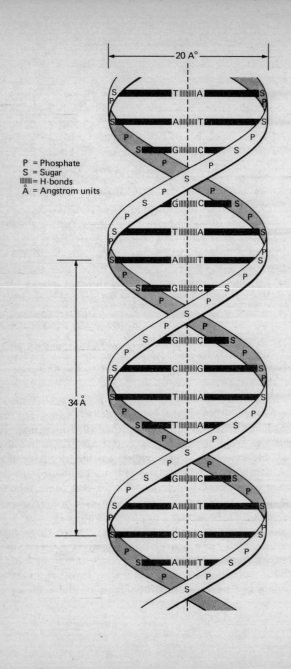

together which were to have major consequences for later studies. In 1938 Astbury and Bell succeeded in obtaining DNA in a crystalline form. This was highly significant: polymers can form crystals only if they have a highly ordered structure. Crystallinity also means that at least potentially DNA could be examined by X-ray techniques—which theoretically are capable of giving a complete structural map of a molecule—although in fact at least ten years were to elapse before X-rays were used with major success.

Then in the late 1940s more light was thrown on the nature of DNA by Chargaff's chemical work. He found four bases in any one sample of DNA, but the odd observation turned out to be that these four seemed to be divided into two pairs, and the two bases in each pair always appeared in equal amounts. Both these observations suggested a highly organized structure for the nucleic acids, and about this time the Nobel laureate American chemist, Linus Pauling, suggested on scanty evidence that the DNA molecule was coiled up upon itself, and that it had a helical structure. In the early 1950s Professor Wilkins and his group at King's College, London, began a systematic X-ray examination of the DNA molecule. Using DNA from different sources, they came to the conclusion that all shared a basic pattern. One of the curious features of this pattern was that although the distances between two bases turned out to be 3.4 Å, there were repeats of the X-ray pattern at 34 Å. The obvious conclusion seemed to be that DNA is indeed coiled up in a helix, with one turn for every ten bases. The results of the Wilkins group suggested yet another idea to Crick and Watson at Cambridge. Suppose the DNA molecule were not a simple helix, but a double one rather like a spiral staircase, with the sugar-phosphate backbone forming the two banisters and the base groups the stairs. We should recall here that by this

Figure 28(b)   The Structure of Nucleic Acids

(b) *The double-helix according to Watson and Crick.*
   A spiral staircase, where bases (H-bonded) are the flat steps and the S–P–S–P– 'backbone', the 'banisters'. The bases are nearly perpendicular to the sugar (S) of the sugar-phosphate 'backbone' (S–P–S–P–).

time the overall chemical composition of DNA was known: a probable backbone of alternating sugar and phosphate molecules' to which are attached four different side groups. These fall into two groups, purines and pyrimidines. There are two purines— guanine (G) and adenine (A), and two pyrimidines—thymine (T) and cytosine (C).

Watson and Crick had one very important fact at their disposal: viewed head-on the DNA molecule has a constant cross section of 20 Å. They proceeded to make models, arranging the known groups in various ways to see if they could get a feasible structure to fit X-ray chemical data. The Watson-Crick DNA model describes a double-stranded structure, where an A side group is always facing a T side group. The C and G groups are arranged in a similar manner, so that definition of one strand of the double DNA helix automatically defines the other strand. Like all first- class theories, the Watson-Crick model has an underlying sim- plicity. To appreciate this fully, we must consider the actual shapes of both the spiral chain and the side groups. Despite the neces- sarily misleading two-dimensional representation of the shapes of atoms and molecules, we must bear in mind that these entities have a very definite three-dimensional volume and that they can link up only at certain defined angles. Thus, once we have made sure that the backbone of DNA is a double helix, if we try to place on it all the side groups known to be there, we have to do so in only one particular way—otherwise they would not fit. It can also be shown that between the side groups facing each other across the helix stretch the so-called hydrogen bonds. These are similar to, but very much weaker than conventional bonds, but because there are a great many of them the total stiffening of the molecule they represent is considerable.

The structure of DNA we have stipulated does appear to give the right answers to the problems of replication and the trans- mission of genetic material. Replication will occur if the two strands of DNA somehow separate; then each strand can act as a template for another strand to form around it. Only one chain of a helix functions as template, and current work seems to indicate that a new chain is synthesized from separate strands as the parent chain 'unzips'. For protein synthesis, we can consider the nature, number, and sequence of side groups to act as a three-dimensional Morse code, containing the necessary instructions.

USE OF ANALOGIES TO EXPLAIN PROTEIN SYNTHESIS

We are very pleased with ourselves nowadays if we can design a machine which, having been fed with the necessary information, can automatically produce some simple article or perhaps make sure that one or two chemical reactions are carried out satisfactorily. Compared with the complexity of chemical manipulation that goes on in the cell when it is synthesizing protein, this is rather elementary child's play. We have to explain not only the very great number of complex chemical reactions taking place, but also the fact that they happen in the correct sequence, as and when required. It is not surprising therefore that, in order to give ourselves a set of rather simple mental images that are easy to manipulate, we often tend to talk about cells as powerhouses, factories, and information storehouses. All these analogies have some value. They group together existing information, allow us a certain bird's eye view, and act as digestive aids to the amount of complex information we have to consider.

In particular, it is often convenient to talk about the nucleic acids as some type of information-containing systems. It is an interesting new departure to regard chemicals not as simple and potentially smelly things which turn litmus blue or red, but rather as specialized bits of computer memory. This way of thinking arose naturally when we began to deal with molecules much larger than the ones usually described in conventional textbooks, and we had therefore to evolve a language which could describe rather simply that because a large molecule with a great variety of side groups could take part in a variety of chemical reactions, it had, so to speak, prior information about these reactions.

Although these concepts have a number of advantages, we must beware of carrying them too far. The greatest danger is to endow large molecules such as the nucleic acids with a personality of their own, and to begin thinking in terms of a molecule doing or wishing to do one thing or another. This, of course, is only the old vital force in a new guise, or what students of philosophy call a teleological argument: that is, something happens because of the required end result. This is a problem that science cannot answer. We can only describe what we believe to be happening and give plausible explanations for the reasons. Thus when we talk about factories, information centres, and molecules doing this, that and another thing, what we really mean is that certain reactions have been observed and that they can be interpreted in certain ways, and no more.

PROTEIN SYNTHESIS—THE FACTORY

When we look at protein synthesis, we should recall that proteins are composed of chains of amino acids, linked in certain specific orders. Thus the essential requisite for the synthesis of any given protein is that particular types and numbers of amino acids should be collected together and then made to link up in the appropriate order. At the same time, we should also bear in mind that, like most manufacturing processes, protein synthesis requires energy, which has to be supplied.

We know that these syntheses occur chiefly outside the nucleus in the main body of the cell, and that certain sub-cellular particles called ribosomes are involved. Also taking part are RNA molecules associated with the ribosomes, sundry proteins, and enzymes. So the information required for protein synthesis, which we postulate to be stored in the DNA, has to be transferred to the protein manufacturing sites, which also have to be supplied with the requisite amino acids and energy.

Our first requisite is therefore a plan, ground map, or template of the protein molecule to be formed. Such a plan would have to specify two factors: the particular amino acids necessary for the construction of a protein and their sequence along the protein's backbone. Once these two factors have been specified, the length, shape, and therefore functional characteristics of the protein also have been determined. The ground plan is contained in the

DNA, and it has therefore to be conveyed somehow from the nucleus of the cell to the ribosomes in the cytoplasm.

It turns out that this is accomplished through the agency of a single-stranded nucleic acid, RNA, that we have seen to be similar to although smaller than DNA. Our current ideas are mainly due to the biochemists Jacob, Monod, and Brenner. They worked with bacteria in which all life processes happen so fast that they can be conveniently observed (cf. page 162).

The RNA molecule transmitting the plan from the DNA to the synthesis site, called, aptly enough, messenger RNA, has to encode the message. In actual terms, the correct RNA molecules has to be formed—perhaps the only true instance where the medium definitely is the message. mRNA formation takes place at the DNA site, as shown by in vitro experiments. If the appropriate RNA components, DNA, RNA-forming enzymes, and sundry other chemicals are mixed, the correct mRNA-like products can be formed. Messenger RNA possesses a further attribute, not shared by other RNAs. It is reasonably unstable, so one assumes that once its function has been performed it can easily be decomposed and its components used in another synthesis.

We can therefore postulate that DNA has acted as a template in the synthesis of an appropriate mRNA, which would then carry the image of the base-sequence information on one of its strands. The RNA has made its way out of the cell nucleus into the cytoplasm. Here it is associated with one or more ribosomes—clusters of ribosomes consisting of up to a hundred or so are known. In general the size of the ribosome seems to be determined by the size of the mRNA and therefore by the proteins that have to be synthesized.

Any particular ribosome consists of two sub-units that can be separated by sedimentation. The units are termed the 30S and 50S particles, where the S refers to an arbitrary unit of size. A large number of different proteins make up each particle and these latter appear to take part in a continuous minuet of separating and re-uniting, although without being broken down and re-synthesized. We shall be looking further into the mechanics of protein synthesis in a short while, but the ribosome structure

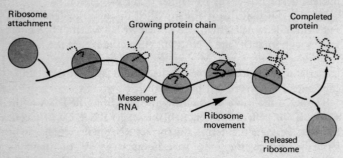

(a) *A simplified view of protein synthesis.*
This is how the genetic code, carried on the messenger RNA, is translated on a ribosome into a protein chain.

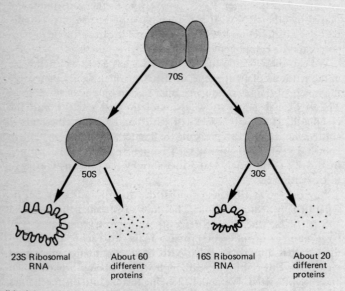

(b) *A bacterial ribosome is dissected into its component parts.*
Figure 29 Protein Synthesis and Ribosomes

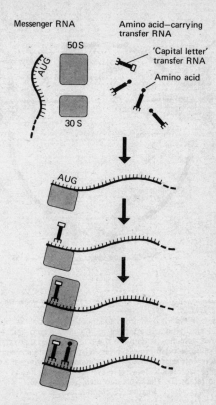

(c) *Early events in the synthesis of a protein.*
The basic components for protein synthesis are shown at the top of the diagram. To begin with, the smaller sub-unit of a ribosome (the 30S particle) becomes attached to the messenger RNA strand, binding to the initiator codon (AUG) which occurs at the beginning of all messages. The special 'capital letter' transfer RNA then joins the complex, followed by the addition of the larger ribosomal sub-unit (the 50S particle) to form a complete ribosome. Finally, the next amino-acid-transfer RNA enters the ribosome, allowing protein synthesis to start.

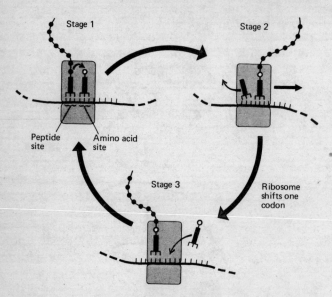

(d) *The new amino-acid (open circle) is added to a growing protein chain (black circles).*

The chain is 'handed' over from the peptide site on the ribosome to the amino-acid site, where it attaches to the new amino-acid (stage 1). The transfer RNA, which was carrying the chain, is then lost (stage 2) and the ribosome shifts along the messenger RNA by a distance equivalent to one triplet codon. This process, known as translocation, brings the growing chain back into the peptide site, leaving the amino-acid site free to receive the next amino-acid (stage 3). The cycle can now begin anew.

Figure 29 (continued)

suggests the presence of specific sites for the acceptance of incoming raw materials, the amino acids, an incoming message site, and a third area where the protein chain can be linked up.

It seems that the first two are located on the smaller sub-unit, while proteins are linked up on the 50S particle. This in turn suggests that the initial step in protein formation is the meeting of incoming raw material and information, in the shape of the mRNA, on the 30S site, followed by interaction with the 50S sub-unit, polypeptide synthesis, and the departure of the finished protein. This theory also proposes a doughnut-shaped ribosome.

We now have our factory, and all we need is raw materials and energy to start and continue processing. The raw materials are the appropriate amino acids which have to be delivered to the factory site. The agency carrying out this task is another RNA, called transfer or tRNA.

Transfer RNAs are rather small on the nucleic acid scale, having molecular weights in the region of 25 000. Although they too are composed of the conventional sugar-phosphate spine and base spine groups, they show some difference from DNA or mRNA. Each tRNA must have an indicator, or recognition site, that will allow it to deposit its amino acid charge at the correct place on the mRNA. These two sites that can be said to recognize each other are called the codon on the mRNA and anti-codon on the tRNA.

It is not clear if the anti-codon site on a tRNA molecule recognizes not only the appropriate mRNA site but also the amino acid required, but for this argument it hardly matters. Some base sequences on tRNAs for the recognition of particular amino acids have been worked out recently and seem to indicate that similar regions exist on different tRNAs—these are sometimes described as loops and consist of between three and five bases. Neither is it clear if a particular amino acid is recognized by only one or by a number of tRNA molecules, although it appears that the latter is the case.

The attachment of the appropriate amino acid to its tRNA is itself an intriguing process. First an activated form of the amino acid is produced from the acid, an enzyme, and ATP. Complex formation takes place on the surface of the enzyme whose specificity ensures that only the appropriate connections are made. The accuracy of this process is marvellous: it has been shown that mistakes occur in only about one case in 10 000.

The tRNA, complete with its amino acid package, now arrives at the protein synthesis site and attaches itself to one of the ribosome sub-units. Its anti-codon, the amino acid site recognition

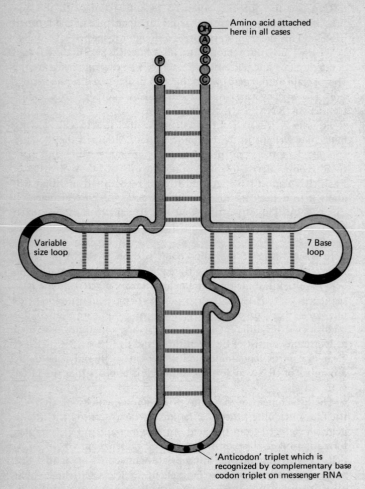

Figure 30   Transfer RNA

This consists of one chain (70–80 nucleotides long) looped on itself and held together by hydrogen bonding between the complementary bases (⏐⏐⏐⏐⏐⏐⏐⏐). Some regions (▬▬▬) appear to be common to all species of transfer RNA.

portion, becomes attached to the codon on the messenger RNA template. The amino acid is deposited and the tRNA, having carried out its function, moves on to an exit site and drops off the ribosomal complex. The ribosomes and mRNA now shift one place relative to each other, the next amino acid arrives and is deposited, and this process is carried on until the complete protein chain is built up. One can visualize this process as the ribosomes moving along the surface of the mRNA containing the template, reading the message it contains, and this is the usual way of describing this aspect of protein synthesis.

The formation of the complex between the amino acid, tRNA, ribosome, mRNA, and the linking of the protein molecule and its final release are all energy-consuming steps. We do not know exactly how the required energy is supplied, although it is realized that both enzymes and GTP, a close relation of ATP, enter into the process.

A further question concerns the relationship between the final product and the ribosome—are different proteins produced by different ribosomes? Recent work by Rich, one of the discoverers of polysomes, the multiple ribosomes, suggests that this is so. Rich isolated polysomes from developing chick muscle at different ages. He could show that these polysomes, when placed in the appropriate environment, were capable of synthesizing muscle proteins and, indeed, could end up with the production of the main contractile elements of muscle fibres. Now it is known that different proteins are synthesized in muscle fibres at different stages of development, so that Rich's work could correlate the size of polysomes found with the protein manufactured. It turned out that the different proteins appearing at different stages of chick muscle development could indeed be correlated with different sizes of polysome.

TRANSMISSION OF CONTROL INSTRUCTIONS IN PROTEIN SYNTHESIS

At this stage we have to introduce the viruses, as a very great deal of the work leading to the elucidation of the role of nucleic acids proceeded with their help. Viruses, organisms even smaller than bacteria, were first discovered in 1892 by the Russian scientist Ivanovski, who found that the juice of infected tobacco plants contained an agent that passed on the infection. These agents were named viruses (poisons) by the Dutchman Beijerinck. Later work also found a group that attack bacteria, the so-called bacteriophages. Of this latter, a number of types destroys the bacterium *E. coli*; they are called the T strains and are numbered T_1 to T_7. In subsequent discussions the T strains of *E. coli* bacteriophages will crop up constantly.

By about the 1930s it became evident that viruses contain only a somewhat primitive control centre: they have either RNA or DNA but not both. Schlesinger, a Hungarian chemist working in London, established that the T_2 bacteriophage is 1000 Å long and is composed of DNA and protein. Later on, others found that the tobacco mosaic virus, the first crystallized virus, contained RNA.

Because they have only a primitive synthetic mechanism, viruses have to use other organisms for their survival and it turns out that these are often bacteria. Hershey and others found that when the T_2 phages invade a bacterium (friend *E. coli*) there is an actual attachment of the virus. The phage content, later shown to be DNA, is emptied into the bacterium and takes over the synthetic mechanism, so that the amino acids and other materials are used to synthesize virus material—in this case the protein coat—but the control comes from the virus rather than from the bacterium. When the synthetic processes are nearly finished, and new viruses made, a special enzyme synthesized on the virus DNA template breaks open the bacterial walls and the newly formed viruses can leave and carry on to infect more bacteria.

The reality of these ideas was shown in an elegant series of experiments based on our ability to mark, or tag, certain atoms. In this case, the atoms involved were the isotopes sulphur-35 and phosphorus-32. Both these atoms happen to be radio-active; therefore it is relatively easy to follow any molecule that incorporates one of them. A further interesting fact turned up; the virus used will incorporate sulphur atoms only into its coat and phosphorus atoms only into its DNA. The technique therefore gives an effective means of marking the outer coat or the DNA of the virus.

If a virus is grown in a medium containing one or more of these radio-active atoms, it will use them in its own metabolism, in the same way it would use an ordinary atom of the same type. So basically the experiment consisted of growing the T_2 bacteriophage, a bacterium-eating virus, in a solution containing either marked sulphur or marked phosphorus atoms. Afterwards, the bacteria and the virus can be separated by centrifugation. It turned out that the bacteria attacked by viruses marked with phosphorus-32 also became radio-active, but those attacked by sulphur-35-labelled viruses were not. The inference was obvious: the infective agent of the virus is its DNA, or part of it, not the virus coat.

That the power of a virus to infect is controlled by its nucleic acid has also been shown by experiments on the tobacco mosaic virus performed by Frankel-Conrat, among others. The protective protein coat of one strain of virus was taken off by means of detergents, leaving free virus RNA. This was combined with the protein coat of a second strain. The infection caused by the newly constructed virus was the same as that given by the first strain: in other words the protein coat had no effect.

Thus experimental evidence accumulated that the control of synthesis resides in the central unit, whether it be RNA or DNA. The next question that had to be answered concerned the way in which a control schedule could be translated into synthetic activity of the cell.

Hershey, Chase, Volkin, and others performed the classical experiments that finally nailed the agent of central control. Again, the tools of the experiment were the bacterium *E. coli* and the T_2 bacteriophage. *E. coli* was grown in a medium containing phosphorus-32 at varying rates, and the incorporated marker showed up in most of its RNA. The next step was to grow *E. coli* in a medium containing labelled phosphorus and add the bacteriophage, which then infected the bacterium. If after infection the bacterium was speedily removed from the solution, it was found that a small fraction of its RNA, less than 1 per cent, had rapidly incorporated the phosphorus-32.

In order for the tagged phosphorus to have been incorporated in the bacterium's RNA, it must have formed very rapidly after infection: one can say that it had a high turnover. Later it was also shown specifically that these rapidly turning over RNA molecules had base sequences complementary to those of the phage, and not to those of the bacterial DNA. The unambiguous conclusion was that after the bacteriophage attacks, something happens at a quick rate that results in the production of bacteriophage-like RNA molecules.

This notion was further elaborated by what are called hybridization experiments. If DNA from the bacteriophage and the newly formed RNA were mixed together, heated, and slowly cooled, complementary strands form that can be picked up by centrifugation. In this way, after the bacterium's own DNA was broken down following infection, the relationship between the bacteriophage's control and the newly formed bacterial RNA could be further supported.

At the same time in France, Jacob and Monod were also conducting an extensive array of experiments concerned with the sexual and feeding habits of bacteria, some of which we shall be considering under the heading of genetic mapping and the operon model. These experiments suggested that it is not sufficient to have DNA or RNA controlling protein synthesis in the presence of the required raw materials. Some agent was also necessary to carry the instructions from the genes to the cytoplasm containing the protein factories. By the late 1950s, this work was sufficiently advanced for Jacob and Monod to predict the existence of a messenger, and in 1961 they found it in the classical experiment by Monod, Jacob, and Brenner.

This experiment relied on tagging portions of the bacillus *E. coli* and devising methods of following the identified pieces. In a simple form, one can regard it as progressing through three distinct steps. First, *E. coli* was grown with carbon-14 and nitrogen-15 so that in the normal course of events the ribosomes of the bacterium were also labelled. Because of differences in weight, ribosomes from the tagged bacteria could be separated from those of normal, or light cultures by using a density gradient—obviously the heavy faction settles lower down in the tube.

The second step involves the consideration of the question: it is known that when *E. coli* is infected with T_2 phage, the protein made is specific to the phage rather than to the bacterium, but in addition, as we have seen, rapidly produced RNA is also present, and was detected by tagging the phosphorus used in its making; is this newly formed RNA a messenger from the phage DNA, or are new ribosomes designed to make phage protein formed in response to the phage attack? In other words, who controls the RNA?

To obtain an answer, tagged, or heavy bacteria were transferred to a normal, or light medium directly after infection by the T_2 phage. Any new RNA produced in response to the infection therefore had two options: it could either associate with the heavy ribosomes, made originally by the heavy, tagged bacilli prior to the infection, or it could get together with the normal, light ribosomes from the light culture, formed after the infection. In order to find which RNA molecules were present before and which after the infection, a further labelling technique, this time dependent on phosphorus-32, was also used.

When both new and old ribosomes and RNAs were sorted out by extraction and density measurements, the three scientists found that all newly formed RNA was associated with the heavy ribosomes, that is, with ribosomes formed before infection. They could also show that new protein, formed under the direction of the phage DNA, was also associating with the heavy ribosomes which were there before infection. The interpretation of this finding was that it was not necessary for new ribosomes to form as a prerequisite of new-style phage protein manufacture. Indeed, the ribosomes can be considered as neutral agents on which the RNA formed makes its imprint. Ribosomal RNAs do not possess the information necessary for the production of protein; this is conveyed to them by the newly formed, phage-oriented and rapidly produced RNA species we now call messenger RNA. Subsequent experiments, mainly by other groups, also found that ribosomes are often working in groups, the so-called polysomes. These experiments, conducted on living bacteria and later in animal cells, were subsequently reproduced in a simplified form in the test tube. These were the classical experiments done in the U.S.A. by Nirenberg and Matthei, which provided us with most of our initial knowledge of the genetic code. The Nirenberg and Matthei experiments showed that it is possible to devise artificial, cell-free systems where amino acids would polymerize into structures similar to, although simpler than the proteins made by the living cell, using synthetic, and therefore infinitely simpler polynucleotides, a child's versions of messenger RNA. Elaboration of these experiments led to the synthetic production of RNA molecules and to the whole area of work usually covered by the blanket and somewhat misleading title of 'life in a test tube'.

The Nirenberg experiments essentially consisted of synthesizing artificial messengers of known composition, that is, containing known base sequences. This could be achieved by using enzymes that are known to join bases by specific linkages: latterly it can also be accomplished through a mixture of chemical and enzyme synthesis. The man-made messengers were then placed in a cell-free system containing all the materials necessary for protein synthesis, including the twenty amino acids known to form proteins. The results were that messengers containing different base sequences synthesized different polypeptides—relatively short chains of amino acids. The known base sequence on the messenger could be correlated with the polypeptide formed by tagging different amino acids in the soup of raw materials and establishing which were incorporated in the polypeptide.

THE GENETIC CODE

The next step is to inquire into the precise meaning of the DNA's instruction, or the exact way in which a number of bases can define an amino acid in a certain sequence. This detailed problem is usually described as the genetic code. There are a number of immediate questions we have to answer. How many bases are necessary on the DNA to define one amino acid? Does one group of bases determine only one amino acid, or more than one, or can there be an overlap? In what form is the instruction given to start and to terminate the sequence? Is any given sequence only a list of instruction? Are there spaces between them corresponding to the punctuation marks in a written message? And perhaps the most important question of all: as most of our knowledge comes from bacteria and viruses, can we assume that processes shown by them are also valid for other forms of life? Is the genetic code universal?

When we consider the complexity of DNA and the other chemical substances taking part in these processes, it becomes evident how difficult it must be to obtain meaningful experimental answers to these questions. It is one of the greatest achievements of contemporary science that evidence has indeed been obtained although not before a great many other disciplines—physics, engineering, information theory among others—have lent their ideas and technical skills for this task. The final, agreed interpretations, especially in a somewhat simplified form, appear relatively simple. But one must remember that behind every statement there is an immense amount of often tedious and frustrating work, and, following the general rule in science, it is only by hindsight that the logical sequence of events becomes clear.

Most of our evidence on the genetic code comes essentially from two broad types of experimental method. In one of them, we disturb the base sequence on a DNA molecule and look for the consequences. In the other we try to simulate what happens in a real cell by synthesizing somewhat simpler but comparable systems.

Under normal conditions a mutation, a mistake in the replicating or transcribing process, occurs once in about 10 000 times, so that even using bacteria we should have a long time to wait for anything to happen. But to cause a mutation, to disturb the base sequence on the DNA, we can use some agency which would do this for us. Such agencies are called mutagens. They can be some type of radiation—X-rays, ultra-violet light—or a chemical. Of the chemicals, there are a number of substances similar to the base side groups normally found in DNA. They link with the normal base through hydrogen bonding, allowing the wrong base to slip into its place. Another of these chemical mutagens is nitrous acid, which acts as an oxidizing agent. Yet another, one of the acridine types, can cause additions and deletions by inserting itself into one chain and binding with the other.

The other method consists essentially of manufacturing an artificial template, normally of the RNA type, and determining whether it can control the assembly of amino acid sequences. For instance, an enzyme called polynucleotide phosphorylase, discovered by the biochemist Ochoa, will promote the formation of long chains from bases. Such sequences are not thought to have an information-carrying structure in the same way as RNA or DNA, since the numbers and sequences of bases are only determined by the amounts and numbers of them that are initially placed in the solution. Yet they are sufficiently similar to the real compounds to allow reasonable deductions to be made about their actions. The synthesis of such polymeric pseudo-RNAs also allows us to make compounds containing only one type of base and to see if they then promote the synthesis of a protein-like structure containing only one type of amino acid. Recently Khorana in the United States succeeded in synthesizing specific RNA sequences using a combination of enzymes and chemical methods.

When we come to the actual coding problem, it is straightforward enough to work out necessary correlations between the four bases present in DNA and RNA and the twenty or so amino acids which have to be incorporated into a protein. If each base is a code for only one amino acid, we could have only four of them. If two bases are a code for an amino acid, they could define sixteen. However, if each code consists of three bases, we have sixty-four signals. Although this would leave forty-four signals for which there is no obvious use, yet it does seem reasonable that groups of three, or multiples of three bases, constitute the coding units on the nucleotides. One can consider three pieces of evidence for this interpretation.

If we introduce one addition or deletion into a DNA sequence, we shall have destroyed the power of the whole sequence. No protein forms. The same effect occurs if there are two deletions or additions. But if there are three changes—deletions or additions—provided they are not too far apart, the sequence can still function.

As shown in Figure 31, these observations can be explained on the basis of triplet coding.

We can also obtain evidence from our simulated systems. If we arrange for the necessary starting materials to be present for the synthesis of protein, but instead of DNA we use a synthetic template consisting of three base sequences, we still obtain amino acid sequences.

We can therefore be reasonably certain that three bases in the DNA determine, through transmission by the mRNA, the amino acid sequences in the proteins manufactured by the cell. We can then inquire further into the details of this genetic code. The first obvious question to ask is if there is a one-to-one correlation between given groups of bases, or is there some overlap between them? And even if there is not, will one amino acid always be determined by the same base group? And in the affirmative, can we establish any proven correlation between a given group of bases and a particular amino acid?

If we make a model of DNA, or the messenger RNA which acts as the template for protein synthesis, it seems unlikely that a group of three bases should be able, by its shape, to determine the

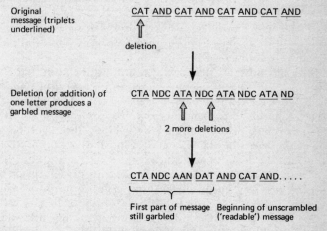

Figure 31   Frameshift Mutations

particular amino acid sequence. They are too far apart. However, when we also consider that the incorporation of amino acids proceeds through a mechanism involving the transfer RNA and the ribosomes, this objection can be overridden.

We can determine whether an induced change in the base sequence changes one or more amino acids in the protein being formed. It turns out that when a mutagen interferes with one base, the effect is shown only in one amino acid. This evidence is further corroborated by work with simulated systems. We have seen that artificial base polymers can act as templates for rather simple amino acid sequences. We can therefore set up our experimental conditions so that only one of these poly-bases is present, say poly-AAG. Under such conditions we get amino acid sequence that can be interpreted as being synthesized on templates of AAG, AGA, or GAA. In other words, although the experiment has not determined where the message starts, it does indicate that one group of bases codes only for one amino acid.

But if one base group always codes for one amino acid, is it possible that one amino acid is not always coded by the same base group? In other words, is the code degenerate? Experiments indicate that it most probably is. If we consider again our simulated, and therefore simplified systems, we find that not only can we obtain correlations between particular bases and amino acids, but we can also work out the base triplets which will code for any given amino acid.

It seems probable that only the first two bases in a triplet sequence carry precise information relating to a particular amino acid, while the third one may be common to a number of amino acids. It has also been found that some base sequences do not code for any amino acid at all, and we interpret their function as signalling where the coding sequence has to start or stop, or where a comma should occur in the message.

The next obvious question is to inquire why the code should be degenerate—why go to all the extra difficulty of introducing seemingly superfluous coding sequences? One must be very careful in interpreting results given in answer to this question, since there is nothing easier than to assume some grand design which, foreseeing future difficulties, had made provision for them from the start, rather like building motorways in the time of horse-drawn traffic. Basically, we rather tend to believe that the initial emergence of genetic systems, which, as we shall see, are virtually universal among all living things, was a result of a series of chance accidents, although later development obviously tended to favour the emergence of forms most suited to contemporary environments.

Therefore the mechanism was not created because of particular functions it has to accomplish, but given a basic framework the particular functions developed to utilize fully a given situation.

A degenerate code has a certain survival value. If, for instance, one base becomes damaged, protein production can switch to an equivalent base sequence and therefore reduce the chances of non-viable offspring being produced. Similarly, some organisms may have a weakness for a particular base, in which case it can be incorporated into sequences necessary for an amino acid which would be required in any case.

The argument about the universality of the genetic code can also proceed along similar lines. An intuitive argument tends to favour the genetic code not being universal. The evolution of species proceeds through mutations, and therefore it seems logical that if in the beginning an organism has, say, two base sequences coding for one amino acid, then as the organism develops into two different species, each sequence will have a preference for one of these sequences only. We have also seen that although in every system A and T, and G and C are present in equal quantities, governed by their ability to take up their positions along the DNA spine, the ratio between A + T and C + G varies from species to species.

It would seem an attractive idea to postulate that the biochemical mechanisms of all living systems work in exactly the same way. This is far from proven as yet, due mainly to the innate complexity of these mechanisms. It seems possible, on the experimental evidence already obtained, that while the outlines of the genetic code are the same wherever we look along the spectrum of life, there are certain differences. One can say with some conviction that the presence of a template will control cell function, and thus the DNA of one species can operate on the cells of another. But the transfer system, consisting of tRNA and associated enzymes, seems to be species specific, which rather spoils ideas about transferring wholesale data from bacteria to the mammalian world.

GENETIC MAPPING

Clarification of how DNA controls protein synthesis through the mediation of messenger RNA raises a further question concerning the relation of genetic information to the DNA ground plan. We have seen that as genetic studies advanced, we have come to associate the handing over of hereditary material with discrete units, the genes, and we came to realize that the genes reside in the chromosomes, eventually identified with DNA. If this is the case, it should also be possible to draw an exact correlation between genes and proteins, in other words to identify the site of the information on the chromosomes with the message passed. Work tending to this end is usually described as genetic mapping. Genetic mapping has other uses too, even if we discount the more than problematical interference with heredity, test-tube babies, and other pseudo-scientific paraphernalia that the science correspondents of the heavy Sunday newspapers tend to use for filling in vacant spaces. Genetic mapping, by allowing us to associate portions of the chromosome with proteins synthesized, would also allow us to obtain a far better idea of important matters connected with biosynthesis. For example, if a number of enzymes are involved in a given synthesis, are their genes situated close to each other, and are they read in any particular order? The number of genetic sites involved would also corroborate ideas on the number of steps (a function of enzymes) necessary for a given metabolic pathway.

Finding the genetic sites depends on techniques similar, as we shall see, to those used in probing the nucleotide control of protein manufacture. Essentially, the work was carried out on very small organisms—bacteria and viruses which, since they were present in very large numbers, could be expected to throw up a reasonable number of mutants within an experimentally manageable time. Mutation could also be encouraged by one of the methods we have already discussed. If, therefore, we have a collection of mutants, we can observe their responses to given external stimuli—in actual practice this involved observing whether different mutants can all grow on a given medium. It is permissible to assume that the inability of a particular mutant to grow means that mutation has knocked out a gene responsible for the production of an enzyme necessary in the processing of one or more of the nutrients present in the culture medium. That is, the lost gene was responsible for one or more steps in the required biosynthetic pathway.

The idea that inherited characteristics are responsible for personality or physical defects in man has, of course, a venerable ancestry, going back all the way to the ancient Greek tragedies. The thick Habsburg lower lip, haemophilia in the Russian royal family, or the recently-argued afflictions in the English royal house in the eighteenth century bear witness that inherited genetic mutations in man very much leave their mark in the circles in which they occur. In an elegant series of studies in 1908 Garrod showed the persistence of certain diseases in families over the generations. The explanations found for these are rather interesting.

Three types of illnesses were involved: (1) alcaptonuria, (2) albinism, and (3) phenylketonurea. Of these, (1) and (3) can lead to mental retardation. Subsequent work showed that all three diseases are connected by missing links in a relatively small metabolic pathway. Alcaptonuria occurs if the enzyme necessary for the breakdown of homogenistic acid into carbon dioxide and water is missing. Homogenistic acid accumulates in the urine, and, when oxidized by atmospheric oxygen, turns urine black. Albinism is due to the blocking of the pathway producing melanin, the skin pigment. And finally phenylketonuria results when the enzyme is lacking for the transformation of phenylalanine to tyrosine.

In 1941 Beadle and Tatum began to publish work that was to establish the truth of the idea that one gene is essentially responsible for carrying hereditary information necessary for the production of one enzyme. Their experimental organism was the red bread mould, *Neurospora crassa*, whose presence can easily be detected.

In the natural state *Neurospora* lives happily enough on a simple diet of biotin (a vitamin), sugar, and some inorganic salts; in other words, it possesses the metabolic mechanisms to convert these nutrients into all the more complicated compounds necessary for its existence. Beadle and Tatum irradiated the mould with X-rays and ultra-violet rays to produce mutations, some of which turned out to be incapable of existing on the natural wild variety's simple fare. However, when the wild variety's minimum diet was enriched by additional material, the mutants could again thrive. The experiments showed that there were materials which allowed the survival of the mutants. They turned out to be simple amino acids and vitamins. It is reasonable to argue, therefore, that something was lacking in the mutants necessary for producing these simple materials. If the synthesis of these essential compounds goes through more than one step, it would be necessary for only the enzyme required for one particular step to be missing; if the organism is supplied with raw material beyond this step it can still carry on. Thus, by varying the enriching materials in the medium, it became possible to trace the various steps required in the biosynthesis of the various amino acids and vitamins and hence argue the existence of the one gene, one enzyme correlation. Subsequent work showed that this idea is essentially correct although there may be a number of additional complications in any given case.

Subsequent work by Tatum and Lederberg in the 1950s showed that the same sort of ideas can also be applied to bacteria. In particular, they showed that when testing for the dependence of a characteristic such as growth on nutritional factors, mutants sometimes reverted to the wild strain and behaved as if the mutation had never happened. The reason for this must be the receipt by the bacterium of genetic material it had lost during mutation, and the recombination of genes has indeed great importance in considering the primitive sexuality of bacteria. We shall be considering bacterial sexuality in the control of protein synthesis by nucleic acids, but we can anticipate by assuming that bacteria can take part in a primitive sort of mating during which one of the pair, the male, inserts a small portion of its nucleic acid into the other, the female partner. This type of behaviour goes under the name of conjugation. By carefully observing end results and varying conditions, it is possible to establish at least a rough genetic map.

Conjugation as a genetic mapping technique has distinct disadvantages. During mating only a portion of the DNA is transferred from one partner to the other, so that information will be lacking about the distal, or remaining portion of the male DNA. To get over this limitation, Lederberg developed an improved method based on his discovery of what is called bacterial transduction. In simple terms, this involves the transfer of genetic material from one bacterium to another, not by conjugation but through the mediation of a bacteriophage.

Lederberg found that portions of genetic material were transferred from one mutant strain to another even when they were separated by sintered glass, and the explanation eventually found was that the agent was a virus. The virus infected one of the strains and in so doing sometimes pinched off a bit of the bacterium's DNA, which then became wrapped in the protein coat of the virus. The virus, being much smaller than the bacterium, could cross the sintered glass barrier, and when in contact with a second strain on the other side of the frontier could deposit the DNA part broken off from the first. In this way a mutant strain lacking a particular gene could sometimes receive it again through the virus intermediary, and would therefore revert to its original, wild state. The parts of DNA carried this way were very small, and it proved possible to correlate the DNA part, the gene, with the effect it caused.

REGULATION OF PROTEIN SYNTHESIS: THE OPERON MODEL

We have considered various parts of the jigsaw puzzle which together make up current knowledge of the role of nucleic acids in the control of protein synthesis and the transmission of genetic information. As with all true science, this knowledge is anything but static; new facts become known, details have to be modified, and sometimes substantial portions of the apple cart are upset. However, in broad lines we have now seen the role of the control centres in cell function.

One very important question remains: Who controls the controllers? Any simple explanation might give the impression that all the systems are going full blast all the time, that on a submicropscopic scale the nucleus and surrounding regions of the cell resemble, if anything, Oxford Street in a permanent rush hour. This would be greatly misleading. As we shall be discussing in this section, there is actually a very fine control mechanism translating the requirements of the cell into action, and no more than the requirements. This series of mechanisms goes under the generic name of the control of protein syntheis, and the model we are likely to regard as accurate is the operon model, developed mainly by the French biochemists Monod and Jacob, although a number of British workers (e.g. Brenner and Hayes) and a galaxy of American ones (e.g. Stent and Lederberg) have also made significant contributions.

We can take as a convenient starting point the observation that bacteria have a sex life, if only of a primitive sort. We have seen that the male of the species can introduce pieces of genetic material into the female, which then becomes capable of carrying out functions such as going on a new diet, of which it has hitherto been incapable. Let us take a strain of our old friend *E. coli* and put it into a medium containing lactose, whereas previously it was used to feeding off galactose. The interesting observation to note is that under these conditions, the bacteria can show two types of behaviour: some can adapt to feeding off lactose in a very short space of time. Others take longer, and this period of time can be correlated with the time taken for them to receive through conjugation bits of DNA from the males. Some, the so-called constitutional mutants, never do manage to adapt themselves to the lactose diet, presumably not having received any DNA parts from their males.

We can turn this observation around and say that the experimental results can be explained if we assume that all the bacteria, except the constitutional mutants, possessed the apparatus to utilize lactose if required, but the apparatus was normally dormant. Another observation also supports this theory: the enzymes produced when the bacteria are feeding off the solution containing lactose are not present when galactose is the sugar in the nutrient.

This is enzymic adaptation. But we can also generalize the idea, later proved by experimental evidence, by saying that in all cells there is a switching mechanism that activates only those functions of protein synthesis which the cell requires at a particular time, and indeed which controls the amount of protein the cell manufactures in response to a stimulus. In general, we can think of this switching action as a series of feedbacks running along a series of interlinked chemical reactions. Although the technique is chemical, the underlying function is still the same as with all feedback mechanisms; once the optimum output is defined, if production falls below the mark the feedback mechanism increases production, and conversely decreases it where there is a danger of too much material being manufactured. Because the bacterial switching system works on interlinked chemical reactions, a series where the production of one substance regulates the production of the following substance, in theory at least any member of the chain may be the regulating agent. The truth of this statement is obvious, if only on logical grounds, since if the synthesis of every product in the chain depends on the availability of its precursor, any stoppage along the line would bring to a halt the production of all succeeding materials.

We can therefore postulate that to account for our experimental findings we can think in terms of two mechanisms, among others, going under the names of enzyme induction and repression. For the purposes of enzyme induction, we stipulate some unspecified mechanism which will stimulate the production of one or more enzymes that the cell needs under a particular set of conditions. Conversely, enzyme repression would stop the production of unnecessary enzymes. Both these mechanisms would operate under particular outside influences, in this case certain chemicals in the food available to the organism.

The immense intellectual jump Jacob and Monod made was to unify these somewhat sketchy ideas into an integral whole under the notion of the regulatory genes, currently abbreviated to the idea of the operon model. Assuming the existence of some type of regulatory mechanism, there can be two general cases. First, it is possible that parts of the control mechanisms are permanently repressed. The repressive mechanism can become deactivated in the presence of specific inducers; in the example we looked at, this would be the lactose sugar, or some compound connected with it. In the second case, it is possible that parts or the whole of the complete control mechanism is permanently in a state of activation by a number of inducers, which are deactivated by some external stimulus. Currently, experimental evidence points in the direction of the first of these options.

Evidence for the permanent existence of repressors comes from mutation experiments. One can obtain mutations of bacteria which first synthesize particular enzymes only in the presence of inducers. However, after the inducer nutrient has been taken away, the bacteria still continue to synthesize these enzymes at a high rate, suggesting that the mutants no longer synthesize the previously present repressor substance. Since mutation implies genetic change, we can also conclude that the original production of the repressor was carried out under genetic control. These mutants, since they have undergone a change in their make-up, are called constitutional mutants.

Further experiments involving bacterial mating also suggest the same conclusion. We have seen that in conjugation only small parts of the male DNA are transferred. Therefore if a female bacterium starts synthesizing a new enzyme after mating, we can be pretty sure that this has come about from genes transferred by the male. The enzymes produced by these new male genes were originally functioning in a cell environment containing the repressors and inhibitors of the male bacterium. Now they are immigrants to the female bacterial environment, which does not contain the home repressors and inducers. What has been observed in a number of cases was that the production of the new enzyme starts off at full rate, as in a constitutional mutant, but it tails off, and after about an hour enzyme production slows to a standstill. These results can be interpreted on the assumption that since the female bacterium has no inducer to trigger enzyme production, there is nothing to keep it going and therefore it stops. Additionally, the male DNA bit transferred to the female can also carry a code for repressor production, which switches off the manufacture of the particular enzyme.

If there is indeed a production of repressor molecules under control of the male DNA, the situation affecting the female bacterium becomes very similar to that of a true mutant, which loses the ability to synthesize a particular enzyme—only in this case instead of losing the gene responsible for its production, the bacterium has its ability to produce cut-off, rather like cutting off all means of delivery from a factory. These experiments lead to the conclusion that there must be genetic material concerned with the production of regulators incorporated in the DNA molecule, which came to be called regulator genes. Distinguished from them are the genes responsible for the production of amino acids necessary for protein, and therefore enzyme synthesis, which now go under the title of structural genes. It is also possible to distinguish between these two types by damaging the regulator genes, when the consequence can be traced on subsequent synthesis by the structural genes.

We have therefore arrived at a broad picture which stipulates a number of structural genes on the DNA sequence. Somewhere in their neighbourhood are the regulator genes, producing repressors which, directly or indirectly, can combine with the structural genes or their products and choke off the manufacture of particular enzymes. The enzymes themselves are responsible for the many chemical reaction sequences that define the way a cell lives and dies. Should the pathway be blocked at any point, all subsequent stages will be absent; conversely the presence of material at any stage will allow the production of succeeding stages.

Refining these experiments brought to light another group of genes which are now called operator genes. These provide the actual switching mechanism for the structural genes belonging to

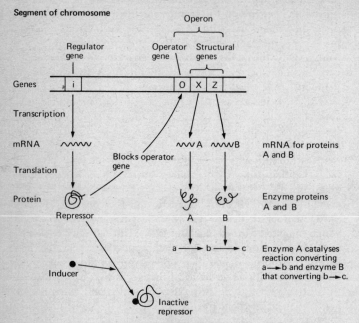

Segment of chromosome

**Figure 32   Operon Model for Control of
Lactose Metabolism**

If mutations occur on regulator (i), a gene which makes a repressor molecule,
or an operator (O), a gene which binds a repressor, synthesis of enzymes A and B
is no longer regulated by the presence or absence of inducer molecules. Mutations
affecting structural genes X or Z will lead to defective enzyme synthesis.

one set. In order that a set of structural genes may operate and
produce a particular enzyme, we have to have a repressor which
is inactivated by an inducer, which in turn switches on the
operator controlling the structural genes. At this stage, all this is
theory.

However, if we set up the model, we can work out various
constellations of conditions and see if they do indeed occur in
practice. One of these models involves the ability of *E. coli* to
produce two types of enzymes for two types of chemical reactions.

The exact nature of these reactions is not important for this argument, so we can call them A and B.

We can now summarize our current ideas on how enzyme synthesis, and by analogy how all protein synthesis, is regulated: how specific, required portions of the DNA molecule are brought to play a part. This we can most clearly do in the form of a dictionary of DNA control. We define as *operon* the cluster of genes, or the genetic unit consisting of adjacent genes, which will typically provide instructions for the synthesis of an enzyme.

The operon is switched on or off by the operator, which turns out to be another cluster of genes near the operon. The operator controls the operon, either directly or indirectly. Our knowledge on this point is not yet conclusive.

We can imagine the operator to be in its inactive state combined with a *repressor*. Repressors are thought to be proteins and in certain cases have been proved to be. Repressors have the added capacity of combining with an *inducer*. If this occurs, the operator switches on, that is, it is no longer repressed. Inducers can therefore be described as molecules which can cause an increased production of an enzyme necessary in their uptake or metabolism.

Taking this terminology a stage further, we can therefore have inducible enzymes and repressible enzymes. Inducible enzymes' rate of production can be increased by the presence of inducers in the cell. For example, we have noted that the presence of lactose in the diet of *E. coli* will increase the production of specific enzymes by the bacteria. In the same manner, repressible enzymes are those whose production decreases when the concentration of repressors rises. Again taking the case of *E. coli*, if we introduce through the diet substances which normally occur at some point in a metabolic cycle, the production of their predecessors decreases. The presence of operon mechanisms makes the cell responsive to a metabolic law of supply and demand and thus confers on it the capacity of adapting to changing external conditions.

Yet the story of genetic control does not end here, for there is new evidence coming to light that in many-celled animals, such as man, the operon model may not be wholly accurate. The operon model is essentially a coarse control. It seems that in many-celled animals there is yet another set of coarse controls, operating a step away from the transcription control of the operon model. Evidence for this statement comes from biochemical studies of tissue development and the effects of hormones, and if proved true, may have important results for our developing ideas on cell differentiation and development. The most exciting chapters in our knowledge of cell control are still to come.

CHAPTER 7

controls and regulators

The dictionary defines the word 'control' as the power to regulate and restrain. In our sense, regulation also has to include the ability to activate or decrease activation. On an overall view, control systems of the biological entity have to utilize the existing pathways and mechanisms by regulating the performance of one or several portions of them. Because all biological systems and the biochemical apparatus they contain are so exceedingly complicated, it is obvious that some very intricate, precise, and reasonably foolproof mechanisms must be available in order for the whole to function harmoniously, or at least adequately.

The nervous system and hormones are usually described as the controllers of the body's biochemical functions, and are indeed often involved with what a more romantic age preferred to discuss in terms of emotions, feelings, and aspirations. Their importance is not diminished in our present view even when we link the sighs of romantic love with increased production of sex hormones. However, before describing what are in a sense extra-systemic controls, one should first of all look at what happens at the molecular level of control. This is the level that all control mechanisms must come to in order to express their functions; whatever their origins, their feedback mechanisms, and their detailed modus operandi, ultimately all control agents must effect some change in the disposition or actions of molecules involved in complex chemical reactions. In other words, the basis of all control is power over what happens in a cell, and it is no greater simplification to argue that cell metabolism is the great criterion on which all control and regulation must be judged.

But cell metabolism is only another way we describe the activities of enzymes, which themselves have to act on a cellular level, and our previous statement is therefore tantamount to inquiring into the factors involved in determining enzyme activity. This we obviously cannot do in any detailed fashion, partly because it would fill several very thick volumes and partly because, even so, quite a lot of the details are still unkown. But we can describe some general considerations and see how the better-known control mechanisms fit into the framework thus gained.

ENZYME CONTROLS

Beyond the complexities of biochemical systems, it is refreshing to find that down-to-earth, common-sense considerations can take us a long way. If we consider the amount of activity, or catalytic function, that an enzyme system is capable of performing in its own cabbage patch, a few simple rules can give us reasonably accurate answers. For in any pathway, what determines the overall rate of metabolism is the speed with which the various intermediate products can be turned into one another through the agency of enzymes. We can visualize such a general situation involving, say, four enzymes, and in order to take our discussion a stage further we have to assume that the activity of each enzyme depends on the presence of the compound it acts on, the so-called substrate. This situation was discussed in more detail in Chapter 4.

The most fundamental control over cell metabolism, given a ready supply of substrate, is therefore the availability of enzymes. Such availability is termed the coarse control since the effects obtained are large and changes in them are only carried out slowly.

One method of coarse control must be the degradation of enzymes involved at different points of the metabolic pathway. The other, which raises more fundamental questions, is that the availability of enzymes must also be the result of their rate of production and this can be traced back directly to the genetic make-up and function of the cell. Such a genetic source would, of course, be liable to all the various control mechanisms we have looked at when discussing the problems of protein synthesis by the cell, and it turns out that such basic control mechanisms may also have an influence on cell metabolism, since in fact enzymes, which control metabolism, are themselves proteins synthesized on ribosomes under the control of DNA-directed messenger RNA synthesis. Experimental evidence is beginning to accumulate that there is such genetic control over metabolic processes, at least in a number of cases.

The other form of regulation of cell metabolism is termed the fine control, and as its name implies, it is a faster and more accurate method. We have evidence that the coarse control, which depends on the regulation of the enzyme supply, is slow. For fine control we must assume that the amount of enzymes does not change, but that some agent is introduced which can interfere with the activity of the enzymes already present. Current research indicates that there are two types of such regulators. One of them interferes with the affinity between the enzyme and its substrate; these we can call isosteric regulators. Isosteric regulators interfere with the enzyme-substrate complex and their action depends on the presence of both enzyme and substrate. The other regulator substance falls into the category of allosteric compounds. These wreck the enzyme directly by interfering with its shape. Since, as we shall be demonstrating later, the action of an enzyme depends on the actual shape of the molecule, anything which alters this configuration will interfere with the catalytic capability of the enzyme.

In addition to these two types of regulators, there are also a number of substances which can modify the formation of an enzyme-substrate complex. In a sense these materials are ancillary to the reactions; they might be cofactors, or coenzymes, or proton carriers similar to those we have seen before. They are necessary to the reactions but are not necessarily specific to them. A deficiency in coenzymes or cofactors means that the reaction they affect cannot take place efficiently, and in this sense their absence or presence does act as a regulator.

A most interesting type of control is effected by the allosteric or deforming materials which act on allosteric enzymes. It seems that these regulators are quite a range of smallish molecules that react with the enzyme, altering its shape. We can visualize this by assuming that the enzyme is composed of two parts and that the intervention of the allosteric regulator alters the relationship between them, thus increasing or decreasing activity. But it is not even necessary to introduce the two-part effect; it is enough to consider some small change among the internal linkages of the enzyme molecule which alters its conformation.

The main point is that such an allosteric molecule need not be at all related to the substrate the particular enzyme is acting upon. This means, for example, that the end products of a long and involved reaction pathway can exercise feedback control in this way over a very early stage of the system. The concept of allosteric behaviour also gives clues to the probable action of hormones which regulate activity from a distance. Allosteric behaviour has been observed in a number of synthetic reaction sequences, fatty acid, nucleotide and amino acid syntheses among them.

PACEMAKER ENZYMES

When we argue that the control of cell metabolism proceeds through enzyme regulation, we leave one large question unanswered: Which enzyme? All enzymes may be equal, but for any particular metabolic pathway some enzymes are very much more equal than others, because they represent the potential location of chemical traffic jams. One can see this quite readily if one considers the steps of the usual step-by-step reaction $A \rightarrow B \rightarrow C \rightarrow P$. The enzymes along this route will transform one intermediate product into another at different rates. For the sake of argument, we can define the activities in such a chain so that the first enzyme has an activity of 100 in some arbitrary unit, the second 50, and so on, until we arrive at a reaction where the enzyme activity has fallen to 10. In other words, this particular intermediate product will be transformed into the succeeding material at one-tenth the rate of the processing rate of a material at the start of the metabolic path. If, therefore, we consider the complete sequence from starting material to product, the total rate of synthesis will only be as fast as the production of that intermediate product that is made at the slowest rate. In the absence of any regulatory mechanism there will be a logjam preceding this stage. The procrastinating point also offers the best opportunity for any regulatory mechanism to interfere with the metabolic path as a whole; since it is already determining the rate of the complete process, any further change of enzyme activity at this point would have immediate repercussions up and down the sequence. For example, should a genetic mutation occur and a slightly defective enzyme be produced, the rate of activity would decrease even further. From the point of view of fine control, the enzyme operating at this point would also provide the opportunity of achieving the maximum effect with the minimum of molecular rearrangement.

The sensitivity to interference of these control enzymes gives us the specific opportunity to track them down. For example, we can take all the enzymes in a sequence—provided we know what they are—and observe their activity under test-tube conditions. Obviously, these will only be approximations of the real thing, but if we find that one of the enzymes works appreciably slower than all the others, we can start being suspicious that the particular enzyme fulfils a control function. In this way, it has been found that one of the enzymes in the glycolysis sequence, phospho-fructokinase, which places a second phosphate group on fructose-6-phosphate, does behave as a bottleneck in the sequence. Because this particular point cannot be bypassed, the control action of phosphofructokinase is obviously important.

Another way in which such control-point or pacemaker enzymes can be tracked down is to determine the concentration of actual intermediates in a piece of tissue. Obviously, if in progressing along the metabolic pathway we suddenly come upon a stage where instead of finding large quantities of intermediates, we only come across very small quantities, something happened along the line—assuming, of course, that the analytical techniques used did not suddenly let us down. By writing down such sequences one can often demonstrate how cross-over points arise. This sort of technique is also used, as we have seen, in the determination of mutations; in these cases a cross-over point would come about because a mutating gene had changed the nature of the enzyme (see Chapter 6).

Inhibition of a pacemaker enzyme, and the consequent piling-up of intermediates, presents itself clearly in changes in blood constitution during starvation. Blood sugar goes down, but its fatty acid complement increases. Since both these materials are funnelled into the Krebs' cycle, the implication is that under these conditions the body is trying to ration out the available glycogen stores and make the most of its available fatty acids. One can duplicate the chemistry of starvation in vitro, at least as far as the picture of the blood is concerned. We then find that glycogen utilization proceeds up to the stage of glucose-6-phosphate and fructose-6-phosphate but the process then stops through lack of the necessary enzyme, phosphofrucokinase. The obvious implication is that increased fatty-acid metabolism interferes with a pacemaker enzyme, in this instance frucktokinase, and therefore slows down a metabolic pathway that has become disadvantageous.

PHYSICAL CONTROL OF METABOLISM: THE MITOCHONDRION

One of the most obvious control mechanisms, whether in biochemistry, traffic jams, or crime waves, is to separate physically the instigator from the place where it can act. Described in scientific terms this concept sounds much more impressive and its details are sufficiently complex to have thwarted our complete understanding of it up to now. An important control mechanism in the sort of reaction we have been considering is the physical separation of the materials required. It is important to appreciate that cells are not little capsules or containers containing a homogenous filling, with everything going on at the same time. Cells are highly complex and internally differentiated structures, provided with a variety of highly selective membranes enclosing specific portions. One obviously important sub-cellular portion is the mitochondrion, and it is evident from the available data that its membrane is fully involved in the control of a number of metabolic pathways.

We can consider, for example, the way in which glucose is transformed to carbon dioxide and water. Initially, glucose is transformed to a twice-phosphorylated sugar derivative, fructose-

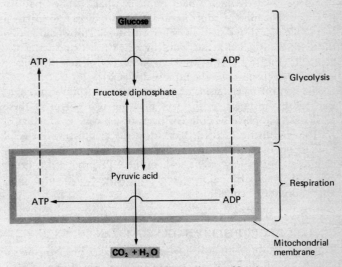

Figure 33   Control of Metabolism by Membranes
Pyruvic acid is formed by glycolytic enzymes outside the mitochondria but is metabolized to CO_2 and H_2O inside the mitochondria. The mitochondrial membrane (shaded area) restricts the movement of ADP and ATP formed during glycolysis and respiration (dotted lines), thus controlling glucose metabolism.

diphosphate, at the expense of two molecules of ATP. Thus, for each molecule of glucose that enters the glycolytic pathway, two molecules of ATP are converted to their diphosphate nucleotide derivative, ADP. The fructose diphosphate is then metabolized to yield two molecules of pyruvic acid. This phase of sugar metabolism takes place in the cell sap. Now, most of the ATP synthesis in the cell occurs in the mitochondria, where pyruvic acid is transformed to carbon dioxide and water, via the enzymes of the Krebs' cycle and of the respiratory chain. If the ATP formed during respiration were freely available to the enzymes which phosphorylate glucose, that is, which initiate the degradation of glucose, and if the ADP formed during this initial phase were freely available to form ATP in the mitochondria, conditions would be created which would result in what is called positive feedback. The more product formed, the more there is to react—a sort of biological 'the rich get richer' situation. This would not do the system any good at all, since by going faster and faster the complete glucose pathway would go out of control, and the glucose stores of the tissue would be rapidly depleted.

The control mechanism, or at least a major control mechanism, in this instance, is the fact that the two halves of these processes take place in different parts of the cell. Production of ADP in the first half is carried out in the cytoplasm, but ATP production in the respiratory chain is a function of the mitochondria. In this way only a little of the ATP produced actually gets to the first part, and instead of positive feedback, one obtains a far more stable condition.

OXIDATIVE PHOSPHORYLATION

The final transformation of ingested foods takes place in the so-called respiratory chain (pages 122 to 128) and provides an interesting example for two types of control mechanisms acting in concert. To recap, the bits coming off the Krebs cycle are stripped of their hydrogen atoms, which in turn are then divided into protons and electrons. Each of these then takes part in a shunt sequence, the end result of which is water. As far as the electrons go, this consists of a number of oxidation-reduction steps, first through NADH, then through the flavoproteins, only to be funnelled into the cytochromes and thence combined with protons and oxygen. The complexities of these steps are renowned, but they do not affect this argument. In essence, we have a slow oxidation-reduction step involving the coenzyme NADH, followed by three faster steps as each of one molecule of ATP is formed from the diphosphate, ADP, and an inorganic phosphate.

What then controls the rate of oxidation? Obviously, if there was no control, it would proceed virtually as far as material became available, and if the energy of the oxidation reaction could not be utilized for synthesis, it would simply be wasted as heat. However, it turns out that what controls these reactions is the availability of ADP and phosphate, which are converted to ATP. Since the free flow of ADP and ATP in the cell is limited, the control operates first of all through the absence of phosphate, which allows no reaction. This state of affairs does assume that the oxidation steps and ATP formation, called phosphorylation, are somehow linked, so that one cannot take place without the other.

There is no doubt that such coupling does exist. The easiest way to demonstrate it is to wreck it. This can be done, and results in a rise in tissue temperature as the energy produced appears as heat, and in the disappearance of ATP, which is no longer produced. This can happen with a number of metabolic poisons ('uncouplers') such as dinitrophenol. Such a condition has been likened to slipping the clutch in a car: the engine revs, but no progress is made.

The question of the coupling mechanism between oxidation-reduction and phosphorylation is far more difficult and, since there is as yet no agreed interpretation of the results, one can only speculate. The most obvious answer is that there must be an actual, physically tangible thing called a coupling agent, which somehow prevents one reaction taking place without the other. The unfortunate part of this hypothesis is that nobody has as yet come across an actual molecule that could be unambiguously identified with the supposed coupling agent.

But perhaps the reason for not finding a coupling agent is that there isn't one, and the second theory of coupling makes a virtue of this necessity. In very simplified summary, the theory suggests that instead of the presence of coupling agents, what happens is that reactants are displaced across the mitochondrial membrane under the stress of osmotic pressure in such a way that the completion of one reaction by its very nature necessitates the start of

the other. Essentially, neither of these theories has yet been proved
or disproved completely and the details of the coupling mechanism
are of no great interest to the non-expert. What is interesting is
that these two mechanisms, coupling and physical availability,
between them control the respiratory rate and therefore the availability of energy for the rest of the organism.

For instance, if we do physical work and therefore our muscles
contract, energy is used up and ATP is converted into ADP.
Another effect will be the increase in the amount of NAD, the
coenzyme carrier necessary for the slow oxidative step. So the
respiratory chain functions accelerate. There is plenty of NAD and
ADP all ready and waiting to be utilized; ATP is formed and is in
turn shorn of its third phosphate in reactions providing the energy
for muscle work. Once the muscle relaxes, ATP and the reduced
form of NAD accumulate; there is less possibility of phosphorylation and the respiratory chain slows down.

HORMONES—SYSTEM CONTROL
IN THE ENVIRONMENT

Hormones are the myth-builders, not only chemical controllers
from one part of the body to another but also messengers from our
biologically not so distant past. When man stalked the primeval
lands he clubbed anything that was small, edible, and moved, and
was in turn hunted down by any creature that turned out to be
more powerful. His fellow animals in their natural state still obey
the same laws. We have progressed; instead of hurling a club we
press a button, instead of running screaming up a tree we drop
unkind remarks beside cocktail glasses. But the underlying biochemical mechanism stays inflexibly the same, and when we
misuse it the consequences are only too apparent.

One group of control agents regulating the organism vis-à-vis
the environment is the hormones, the other is the nerves. Both
systems are large, complex, partly under our conscious control and
partly outside it. In both cases large areas of ignorance remain,
partly filled with the explanation of the moment, only to give way
to some new point of view. Both systems have inspired a mixture

of myth, superstitious fear, and the possibility of a pat explanation of some aspect of our being and essence. Hormones in particular have proved a fertile hunting ground for man's explanation of himself in terms of something tangible and at least theoretically explainable. At some stage or another they provided the illusion of explanations for features of character or the sure way to preserve eternal youth. In both these pursuits they proved to be inconstant agents, but chemically they nevertheless are turning out to be fascinating entities with effects ranging all the way from the physical nature of the subconscious to the prowess of the physical, intellectual, or sexual athlete.

There are a number of specialized tissues in the body, often small nut- or corn-shaped bundles, which secrete a range of chemicals called hormones. Like most definitions this can be elaborated: not all hormones are secreted by these bundles called endocrine glands, and even when they are, they can be purveyed either through a duct or directly through the blood stream. However, for the purpose of this discussion we can regard hormones as relatively small molecules—small proteins, peptides, amino acid derivatives such as adrenalin, or a class of chemicals called steroids, which are secreted by specialized tissues directly into the blood stream. Their main distinguishing feature appears to be that they are designed to act on tissue which may be a long way from their original source, and at the target tissue to interfere with some biochemical activity. This they may do directly or, as happens in a number of cases, one hormone will affect the production of another and in this way a large and detailed cascade builds up with effects all over the organism. Most of the hormones made by the animal organism have been identified and a number also synthesized in the laboratory, yet their precise mode of action at the biochemical level is still, to a great extent, obscure.

The very idea of chemical messengers circulating in the body started with the French biologist Claude Bernard, who noticed that the liver helps to maintain a constant level of sugar in the blood by supplying glucose from its glycogen stores when the intake of external sugar falls below a predetermined level. Although the liver is not a gland, the idea had effect. By 1849 Berthold found that the testicles of cocks could be re-grafted on different points of the bird's body and still provide the accustomed male character-istics—aggression and growth of the comb.

The cock–testicle experiment also opened the doors to real as well as quack ideas of preserving vigour and/or youth. One of the early experimenters was Brown-Séquard, who injected himself at the age of 72 with the extract of guinea pig testicles and claimed to have felt much better. Similar experiments were reported only a few years ago from Roumania, although here the miracle drug was somewhat more sophisticated. Alas, all that these experiments prove is that when very old people are well taken care of and are placed in a situation which gives more interest to life they show improved health and alertness, irrespective of whether the drug is pig testicles, brain extract, or the perhaps less hygienic ministra-tions of a well-qualified witch doctor.

The first real hormone experiment, and the coining of the word itself, is credited to Bayliss and Starling, two physiologists at University College, London. Those were the days of the Pavlovian reflex experiments showing the nervous control over physiological function. Pavlov showed that if hydrochloric acid is present in the small intestine, which it is when food is being digested, nerves running to the pancreas stimulate the excretion of enzymes. These are ducted into the intestine, and as we have seen, have an im-portant role in digestion. Bayliss and Starling severed the nerves reaching into the pancreas and found that nonetheless the sub-sequent insertion of hydrochloric acid into the small intestine triggered off the production of the digestive enzymes called secretin. Therefore, in the absence of nerve messages a chemical agent, christened hormone, must have been present. Later on, it was found that the pancreas secretes two types of chemicals: the digestive enzymes through a duct into the small intestine and insulin directly into the blood stream. Thus the pancreas acts as a ducted or exocrine gland as far as the enzymes are concerned, and in this it resembles other glands of the same type, such as the sweat glands and the salivary glands. An area of the pancreas, called the cells of Langerhans, acts as a ductless or endocrine gland, supplying the blood directly.

The role of insulin, which was discovered in 1922 by Banting and Best, in maintaining the body's sugar balance and therefore its importance in the treatment of diabetics has become very well known. Banting and Best found insulin in the dog pancreas; they discovered that if the pancreatic duct supplying enzymes to the intestine was tied off, not only did the enzyme supply stop but the blood sugar level also fell. The interpretation suggested that the enzymes, mainly trypsin, had therefore the secondary effect of destroying the factor which would have normally lowered the blood sugar level by accelerating its uptake in the tissues. Their experimental technique consisted of tying off the pancreatic duct, waiting for six to eight weeks, and then removing the pancreas and extracting material that would reverse the appearance of sugar in the urine of dogs which had had the pancreas removed. Their success, and the dramatic way in which the life of a patient suffering from severe diabetic disorders was saved as a result, has become medical history.

Since that time a great number of hormones have been discovered and their physiological action elucidated. The most important question from the biochemical point of view is how they carry out their functions, and this we shall be looking at in the next sections.

SEX HORMONES

Hormones active in both the male and the female are secreted by the pituitary gland. In males the end result is the secretion of androgens in the testes. They control secondary sexual characteristics: deep voice, hairy chest, and so forth. Two of the pituitary hormones, the follicle-stimulating hormones, FSH, and the interstitially cell-stimulating hormones, ICSH, are required in the maturation and production of sperm.

But males are far less interesting biochemically, mainly because they are creatures of circumstance rather than nice, steady, cyclical people; this is the main factor in the extreme difficulty of developing a male contraceptive pill. Although the female sex apparatus is much more complicated, it is at least predictable.

Table 3

HORMONES

Source	Name of Hormone	Main Functions
hypothalamus	oxytocin	contraction of uterus ; sperm motility ; milk ejection
	vasopressin	constriction of blood vessels ; retention of water (= antidiuretic)
	'releasing factors'	small polypeptides which stimulate release of hormones from target glands
pituitary gland		
posterior part	oxytocin and vasopressin (derived from hypothalamus)	see above
anterior part	adrenocorticotrophic hormone or ACTH	'stress' hormone : increases synthesis of steroid hormones in adrenal cortex and their secretion
	growth hormone	promotes protein synthesis and growth of various tissues. Excess may lead to gigantism in the young and to acromegaly in adults
	thyrotropic hormone	stimulates hormone secretion by thyroid gland
	gonadotrophic hormones * (ICSH, LH, FSH)	stimulates gonads
thyroid gland	thyroxin and tri-iodothyronine	iodine-containing derivatives of amino acid, tyrosine. Required for physical and mental growth
parathyroid gland	parathormone	controls calcium utilization
pancreas	insulin	controls sugar uptake by tissues ; stimulates fat and protein synthesis

* These hormones are dealt with in detail on pages 193–99.

Source	Name of Hormone	Main Functions
adrenal gland		
cortex	cortisone (a steroid)	affects glycogen and fat metabolism; controls enzyme synthesis
	aldosterone (also a steroid)	controls salt balance (sodium and potassium)
medulla	adrenalin	increases heart rate; promotes formation of sugar from glycogen and fatty acids from storage fats
ovary	oestrogens	responsible for appearance of female secondary sexual traits; stimulates protein synthesis in many tissues
	androgens	responsible for appearance of secondary sexual traits in the male; also promotes synthesis of proteins

At the beginning of the menstrual cycle, the pituitary gland sends out the follicle-stimulating hormone which acts on the follicle, the envelopes containing the eggs maturing in the Fallopian tubes. Under the FSH influence the follicle starts to grow and in turn secretes other hormones, the oestrogens. The main job for the oestrogens is to prepare the womb, or uterus, for the reception of the fertilized egg; under the influence of oestrogens the uterus walls thicken and become well supplied with blood vessels.

Thus, first we have an increase in FSH production by the pituitary; as the oestrogen concentration increases, FSH is cut back, but in turn the pituitary begins to broadcast another hormone, the luteinizing hormone (LH), identical to the male ICSH. The luteinizing hormone increases follicle growth and, at the correct moment, arranges the separation of egg and its enveloping follicle. At this point the follicle starts to turn into a different tissue, called, from its colour, the corpus luteum. The corpus luteum, the yellow body, secretes the progesterone hormones which stimulate the uterus to secrete nutrients for the free-swimming embryo and later help with its implantation and the formation of the placenta. The placenta itself, the lifeline of the embryo, continues secreting progesterone. This controls, and

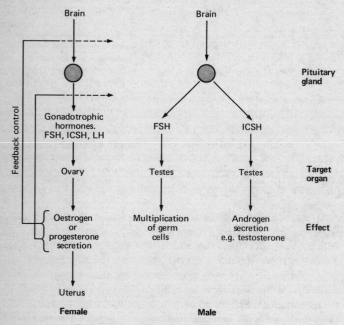

Figure 34 Hormone Control in the Male and Female
Note the two feedback loops controlling the hormone secretion in the female.

stops, further production of hormones from the pituitary, and thus the cycle stops. All this can be followed by noticing the appearance, concentration, and disappearance of the hormones and noting that the appearance of one provides a control mechanism to stop the production of the previous one.

Under natural conditions the aim of all this is obviously the production of the young, and the biochemical mechanism is geared up to it. If through accident or one of a great number of causes fertilization does not occur, the corpus luteum degenerates, the placenta comes away, and menstruation occurs.

The basis of the several varieties of birth control pills involves the taking of oestrogens and progesterone. The hormones thus taken reach the pituitary gland which is so programmed that, given a high enough concentration of these two, the pituitary gland's own production of FSH and ICHS stops. But these latter are needed to ripen the follicle and start its transformation into the corpus luteum, which in turn is the precursor of the placenta's formation. Thus the fertilized egg never becomes established and contraception results. On the other hand, since oestrogen and progesterone are continually in the system at high concentration, the uterus is perpetually being prepared for pregnancy and undergoes the secondary changes, such as the production of muscle enzymes required by this state. This does not appear to do any harm, except that some individuals have claimed to have suffered some of the inconveniences of early pregnancy such as nausea.

These side effects are being overcome by various small adjustments in the usage of the drugs, and up to now no deleterious side effects have been proven conclusively against oral contraceptives. There appear to be statistical correlations between contraceptive pill usage and some deleterious side effects such as blood clot formation. Such possibilities signal caution but certainly do not merit panic reactions. There is one area that has been little investigated and which, one can argue, leaves a trace of suspicion regarding the long-term effects. As we shall argue, hormone control is a two-way process—not only is there control by the hormone at the target tissue, there is also a feedback regulating the hormone supply, a consequence of the tissue control itself. This feedback directly or indirectly reaches the brain and makes its effects felt. It seems reasonable to assume that large quantities of hormones taken into the system must have a feedback to the brain which quite possibly has no effect whatsoever. On the other hand it might have. At this stage we simply do not know.

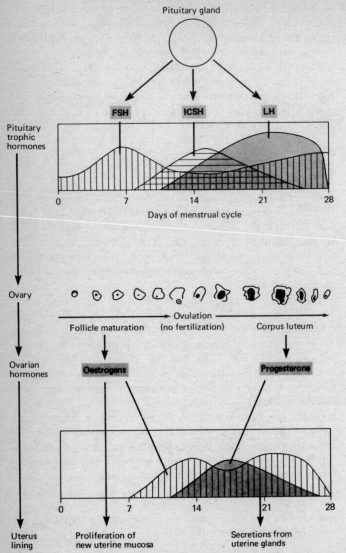

Pituitary gland

FSH   ICSH   LH

Pituitary trophic hormones

0   7   14   21   28

Days of menstrual cycle

Ovary

Follicle maturation   Ovulation (no fertilization)   Corpus luteum

Ovarian hormones

Oestrogens   Progesterone

0   7   14   21   28

Uterus lining

Proliferation of new uterine mucosa   Secretions from uterine glands

Cyclical behaviour in the female, carried out as we have seen under the control of a system of hormone feedback regulators, appears to be established very early on. Indeed, it seems that every baby female is theoretically capable of fulfilling its sexual role from the very beginning and that only some missing stimulus prevents the cycle starting until after puberty. This is also the case with males, although their hormonal secretions are non-cyclic. The stimulus appears to come from a part of the brain called the hypothalamus and may be a response to oestrogens. The existence of the biochemical sexual apparatus in the newly born, obviously a genetically determined state of affairs, led to the suggestion that although the ground plans are ready at that stage, some further imprinting, a sort of biochemical brainwashing, is necessary before it becomes fully effective. Thus, if a newly born female rat is injected with the male hormone testerone, when it grows up it will not show the usual female cyclic pattern. If, on the other hand, a castrated male rat has a shot of female hormone, it will exhibit cyclical behaviour. There is, therefore, a suggestion that potentially we all start off as females, and only a subsequent hormonal imprint determines the biochemical pattern of sexuality.

Figure 35 The Hormonal Control of the Menstrual Cycle

Three trophic hormones are secreted by the pituitary gland to act on the ovaries to promote (1) follicle maturation (FSH), (2) ovulation (ICSH), and (3) growth of the corpus luteum (LH). As a result, these 'target' tissues secrete a second set of hormones, oestrogens, and progesterone, which act on the uterus to 'prime' it for implantation of the fertilized egg. If the egg is not fertilized, progesterone secretion diminishes as LH production falls, the corpus luteum degenerates, and the uterine lining is shed (menstruation). When fertilization occurs, blood levels of oestrogens and progesterone remain high throughout pregnancy and shut off FSH production from the pituitary.

FSH = Follicle-Stimulating Hormone
ICSH = Interstitial Cell-Stimulating Hormone
LH = Luteinizing Hormone

Graphs represent the levels of hormones as a function of time during the menstrual cycle in the absence of fertilization.

FEEDBACK BETWEEN GLAND AND BRAIN

The feedback between hormone-producing tissues is only a part of the complete system of hormone production control. An equally or possibly more important part is taken by the nervous system and its central part, the brain. The brain is connected to the pituitary gland in a number of ways; hormone secretions of the pituitary influence others which in turn exercise control over the brain functions.

The pituitary gland, which sits at the base of the brain, is composed of two main parts. One, the anterior pituitary, is connected to mid-brain structures (for example, the hypothalamus) via special blood vessels, while the other, or posterior part, receives hormones directly from nerves coming from the mid-brain. The posterior part, which is in fact made of nerve tissue, secretes two important hormones, vasopressin and oxytocin, directly into the main circulation. These hormones are derived from the hypothalamus, where they are originally manufactured and stored. The direct release of hormones from nerve-endings is called neurosecretion; the nerve itself is acting as a kind of endocrine gland. The anterior pituitary releases a number of trophic hormones, which act on target tissues such as the adrenal gland, the thyroid, and the gonads, which in turn secrete further hormones to act on target issues. The amount of these hormones in the blood stream acts as a signal to control the release of the corresponding trophic hormones from the pituitary. This type of control can be exerted either on the pituitary itself or on nervous structures in the mid-brain which themselves control pituitary activity.

In addition there is also a special nerve connection linking the hypothalamus to the adrenal gland, another of the general control glands. The adrenal gland in turn is also affected by the pituitary. In this manner there are both hormone and nerve connections between the principal glands and the nervous system and, if the whole vast arrangement is worked out in detail, its most striking feature is its similarity to the elaborate electronic control systems used for highly automated production processes, complete with in-line control and feedback loops.

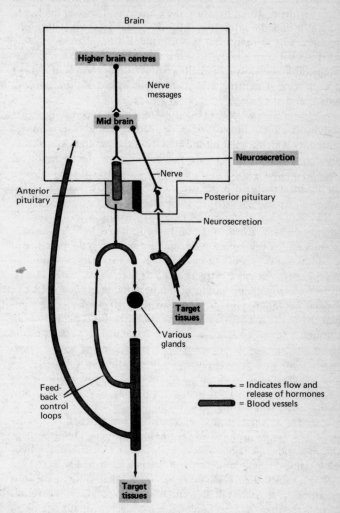

Figure 36 The Relation of Glands to the Brain

The existence of these control loops can also be demonstrated by experiment. If, for instance, the connection between the hypothalamus and the pituitary is cut, the gland together with all its satellites will atrophy. But if the gland is transplanted to some other region of the body, it will retain its power to secrete hormones, although only to a small extent. Without control, the gland is useless, and control is demonstrably linked to the nervous system and the brain.

A great deal of work has been done to show that certainly sexual activity is closely linked to the brain and that the secretion of hormones which trigger off the activity of the pituitary starts from the hypothalamus. It is also suggested that the secretion of oestrogens, much lower down the cycle, can feed back to influence the brain activity; if drugs are used to block the activity of the nervous system, ovulation stops.

The influence or control of external stimuli on sexual activity can also be demonstrated. In birds, for instance, light is necessary to start off the egg-laying cycle, and group effects, the presence of a number of others of the same species, is necessary to stimulate sex life. An even more extreme case has been shown with rabbits; diffuse electrical stimulation of the head or the spinal cord can induce ovulation and pseudopregnancy. If electrodes are implanted in the hypothalamus of the rabbit and it is made to walk through a magnetic field, an induced current is generated in the electrodes and stimulates sexual behaviour. Similar results—both to stimulate and depress—can also be obtained with drugs.

The inference from all these experiments seems to be that direct sensory perception, sight, sound, smell, may have an important influence in stimulating sexual activity; nor does it have to be assumed that these stimuli necessarily have to go through the consciousness, as in the case of the human animal. This is perhaps the most obvious region of existence where biochemistry, physiology, and the emotions meet and become inextricably mixed up. All three influence each other so much that in a sense it is meaningless to try to talk in terms of only one of them. There was a period when it was fashionable to decry emotions and to attribute their effects purely to chemical or nervous control.

Unfortunately for such a logical state of affairs, a description of this kind is incomplete.

Sex hormones are not the only ones to be controlled by brain level regulation. Vasopressin, the antidiuretic hormone, accelerates water reabsorption in the kidneys and therefore decreases the necessity to urinate. The production of vasopressin is also controlled by the pituitary, under the influence of both blood pressure and nervous regulation. These work in the obvious direction that under stress conditions of moderate severity, time which could be usefully exploited in fight or flight should not be wasted in urinating. Similar mild shock conditions also exist after a minor operation or a not too exhausting muscular activity.

A rather interesting example of a limited feedback mechanism occurs in the suckling of mammals. This involves principally two hormones, prolactin and oxytocin, both under the general control of the pituitary gland. Prolactin excites the mammary tissues into milk production and is therefore connected into the complex biochemical circuit attendant on parturition and the feeding of the young. However, prolactin is only concerned with the production of the milk and not its ejection. In parenthesis, it should be realized that breast feeding is not a mechanized version of bottle feeding; under stimulation by the baby animal the milk shoots out and, despite the name of the exercise, does not have to be pumped out. Neither is there any connection between the size of the female breast and the lactating ability of its owner. It is not fat but milk-producing tissue which determines lactating ability and this tissue only comes into action after the birth of the offspring.

When the nipple is suckled, a nerve-connected reflex comes into operation, stimulating the production of oxytocin, which in turn makes the muscle tissue of the breast contract and therefore eject the milk. Curiously, oxytocin has an additional function earlier on in the sexual cycle. Its production appears to be stimulated by coitus and its effect under these circumstances is to induce contractions of the uterus, thus helping sperm mobility. This uterus-contracting effect also comes into operation during labour.

HORMONE REGULATION

To take a more intimate look at the biochemical significance of hormone regulation, we shall consider some of the products of two major glands, the adrenal and the thyroid gland.

The adrenal gland is in reality two closely connected entities, each with its own production programme: the adrenal medulla, which is concerned with emergencies, and the adrenal cortex, a more equable gland primarily involved in the body's salt balance, sugar-protein metabolism, and other matters. Both these glands are activated by a control hormone going under the name of ACTH secreted by the pituitary, and therefore controlled indirectly by the hypothalamus. As we have also noted, there are direct nerve connections to the brain to complete, or complicate, the circuit.

The general role of adrenalin, secreted by the adrenal medulla, is to prepare the body for an emergency, which under natural conditions would result in fight or flight. It is rather interesting to work out the requirements of such a condition. First, it is necessary to have a good look at what is coming: the pupils are therefore dilated to accommodate all visual input. Quick reactions to any movement or noise are also necessary: muscles have therefore to be tensed. If there are ears to be stiffened, so much the better. There also has to be a ready supply of fuel for any bodily exertion that might follow; the breakdown of glucose has therefore to be brought to a high level. Finally, non-essential body functions —for example, digestion—have to be stopped until more relaxed times.

Essentially, the fulfilment of all these requirements is the function of adrenalin. It can be argued that in our culture adrenalin is an atavistic hormone, a relic of a bygone age, probably responsible for a great deal of human unhappiness or confusion, since for better or worse we allow a full expression of its effects only under especially nasty conditions. But even the synthesis of adrenalin is a complex matter. It starts off with an amino acid by the name of phenylalanine and through sundry reactions becomes a compound very similar to adrenalin, called nor-adrenalin. The first step in this process requires an enzyme, phenylalanine hydroxylase, lacking in those suffering from phenylketonuria, a genetic disease. But the production of this enzyme is related to that of a number of substances affecting nervous activity, so once again we have an indication of the connections between nerve functions, enzymes, and hormones.

In more specific terms, adrenalin affects the breakdown of the glycogen stored in muscles to glucose. Glucose is burned, providing a supply of ATP, needed in contracting the muscles. The way in which this role is fulfilled is through the control of the enzyme phosphorylase, a pacemaker enzyme in the breakdown of glycogen. The hormone activates the enzyme through the formation of a complex containing the metal magnesium and ATP. This is a relatively simple reaction and therefore can be accomplished quickly. The intervention of adrenalin has a fast and accurate effect on sugar production and therefore on the level of blood sugar.

An odd use of adrenalin is to inject it into animals before slaughter; increased glycolysis empties the glycogen store of the muscles by conversion into sugar and transports it into the blood stream. Otherwise the glycogen in muscle would break down by bacterial action, producing sugars and lactic acid and hastening the putrefaction of the meat. The treated meat is irradiated by ultra-violet light to kill the bacteria that would otherwise enjoy the sugar produced.

The complexities of adrenalin action can be followed further. The ATP taking part in the activation of phosphorylase has itself to undergo a preceding transformation into a cyclic compound by the name of cyclic adenosine monophosphate, cyclic AMP. This transformation is carried through under the influence of yet another enzyme, adenylate cyclase, which in turn is triggered off by adrenalin. But cyclic AMP does not act directly on phosphorylase either; instead, it activates another enzyme, a kinase, which places the required phosphate on the main pacemaker enzyme, phosphorylase.

On the molecular scale, phosphorylase can exist in two forms. One is the inactive form, and we can visualize it by thinking of two pairs of identical units, each unit having a hydroxyl group attached. In the wings, four phosphate groups are waiting, currently attached to ATP molecules. The action of the kinase is to bring the two pairs of units together and transfer phosphates. As we have seen, kinase is made active and capable of carrying out this role through the intervention of cyclic AMP, which became

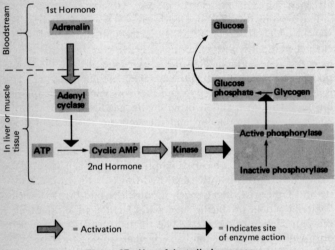

Figure 37 How Adrenalin Increases Blood Sugar Levels

Adrenalin, the primary hormone, activates an enzyme on the cell membrane which converts ATP to a secondary tissue hormone, cyclic AMP. This substance acts on an enzyme which breaks glycogen to glucose phosphate. Finally, free glucose is formed via other enzymes and appears in the blood stream.

cyclic through the mediation of adenyl cyclase, which in turn became activated through adrenalin.

We can describe this chain of events in another way by suggesting that each step of the sequence reflects the allosteric behaviour of enzymes. In this case also we have a small molecule which on the face of it has nothing to do with the main reaction that causes the activation of phosphorylase to increase glycogen breakdown. Yet this insignificant entity changes the shape of an activator, ATP, which then changes the shape of an enzyme, and through this change of shape the further course of events progresses.

It is possible to generalize such observations. The chain of events starts with a relatively small molecule that acts on a large one, generally an enzyme (a protein), by changing its shape. Another way of conveying the same information is to say that the small molecule attaches itself to the allosteric sites on the protein. The changed enzyme then acts on its substrate. The products act as signals to some further event farther along the reaction path. This mode of behaviour is analogous to that of a relay—mechanical or electronic—in an electrical circuit, where a small electrical charge provides the signal for a much larger one, again through some physical transposition of either electrons or large chunks of matter, and the resulting large current can then act at a distant location.

This analysis can be taken one step further. Hormone-regulated enzymic reactions take place on the surface of membranes and therefore membrane shape must have a profound influence on their course and possibility. It can therefore be argued that what hormones, and drugs, and probably nerves, do is to change the shape of the membrane support, and to induce conformational changes to act out their regulating functions. Echoes of this idea are also current in our present views on metabolic changes induced by hormones other than adrenalin, and even on nerve conduction. But if these ideas are correct, they must have some bearing on the general relationship between cells, since these are also influenced by the membranes between them. This leads directly into the as yet mainly unknown world of cell differentiation and the behaviour of cells in groups.

THE ADRENAL CORTEX

The second part of the adrenal gland is the adrenal cortex, under the general regulation of the pituitary gland. The variety of its production is prodigious, consisting mainly of a range of steroid hormones, called the corticoids, which act on the salt metabolism and the protein-carbohydrate mechanism of the body. There is also some manufacture of sex hormones, both male and female, since the steroids can provide precursors. If the gland is diseased, as for example in tumour growth, overproduction of the sex hormones can take place and women can acquire secondary male characteristics, a process called masculinization. Possibly more important is the salt balance, the main function of the steroid aldosterone. As we shall see, nerve action depends on differential transfer of sodium and potassium ions across the nerve boundary, so that when a nerve fibre becomes excited sodium is passed in and potassium migrates out. If there is not enough aldosterone to regulate this function, the ion migration gets out of hand, the muscles stimulated by the nerves become overexcited and, if the condition goes far enough, the patient dies of heart failure.

The other steroid hormones are mainly concerned with regulating the glycogen reserves in the liver and the regulation of protein balance. Protein balance can be affected through the breakdown of protein that, by passage through the Krebs cycle, can eventually be transformed into carbohydrates. Interference comes therefore on the catabolic side, through the enzyme transaminases and amino acid oxidases. These two are in turn regulated by the gluco-corticoid hormones, mainly cortisone. The outline of their action is that they speed the conversion of amino acids into keto acids that can be used in glucose synthesis.

In outline, we have therefore the usual system of a hormone acting upon the production of an enzyme which in turn increases a step in the metabolic cycle. The interesting observation is that this control mechanism is apparently linked with gene action, so that the whole process can, at least theoretically, be thrown back to the very grass roots of cell control. If one injects an experimental animal with certain amino acids, its production of transaminase will increase: the body is coping with the extra load imposed on it in the same way that bacteria will produce the necessary enzymes to adapt to a new nutrient. In theory, there is the same type of adaptive mechanism, presumably through DNA-RNA

control, only in this case genetic regulation operates on the
enzymes through hormone production, rather than directly.
Unfortunately there is also evidence that even this explanation is
incomplete, so that further factors have to be considered.

A set of recent observations suggest even more complex implications for the role of hormone regulation. The response of increased enzyme production under hormone control for some
adaptive enzymes operates only in the grown animal and is absent
in the young. If, however, a very young animal receives an
injection of the inducer substance, in this case an amino acid,
then although the animal's basic enzyme production will only
be little affected in later life, it will no longer react to the
stimulus of hormones or the presence of the inducer. In other
words if the normal adult animal is presented with an inducer,
its enzyme production shoots up, but the grown-up animal
which has been injected shortly after birth shows no appreciable
response.

This behaviour is quite analogous to our usual immune behaviour—our bodies react violently against foreign protein, as
shown in the cases of tissue rejections, but are quite content to co-
exist with all the different proteins in the body. How can they
distinguish? Can we postulate some sort of uniform that all proteins of a given individual organism wear, so that they do not
provoke an immune reaction to their own tissues? The answer
seems to be that it is familiarity, rather than any more specific
mechanism that produces this tolerance. The proteins and other
materials that have been in the organism from birth have somehow created for themselves an atmosphere of acceptance through
the very fact of their presence during this period.

If we accept that this argument is anywhere near correct, we
have opened the door to a whole array of speculation, none of
which can be proven, but all of which present intriguing possibilities. We have seen, for example, that an early injection of a sex
hormone can considerably alter an animal's sexual behaviour. We
suggested that this imprinting possibly operates through the brain
across the numerous feedback loops that hormones possess. It
would be reasonable to argue that early exposure either to excess

hormones or to some inducers would also alter the way in which the individual's brain reacts to subsequent stimuli and, at least logically, it is not difficult to argue that all of us therefore are carrying around the results of a very early biochemical brainwashing. We may call the results the subconscious, but it is in fact some early alteration in our nervous apparatus brought about by chemical means.

A further strand in this argument is the known fact that the foetus receives an overdose of sex hormones and it is quite usual for babies to be born with genitalia out of all proportion to their size. This is tantamount to an injection of hormones shortly after birth; in other words all of us partake in this imprinting reaction. This line of thought can be carried a good deal further, to touch upon the nature of free will and the philosophical and social systems that men have elected to follow. None of it can at present be supported by meaningful experimental evidence, despite the futile attempts of scientific doom-mongers and the prophets of tragedies to come through the results of science. But it is perhaps important to consider the existence of connections between our biochemical, physiological, psychological, and all other systems in order to gain, if not a better understanding—this is at present out of reach of all of us—at least a leap of the imagination in making the necessary gestures of tolerance and sympathy, because free will begins to look very limited indeed when set beside the exigencies of biochemical accident.

THE THYROID

The thyroid is a small gland at the base of the throat which secretes an iodine-containing derivative of an amino acid and provides an important control of growth and speed of living. It was one of the first endocrine glands to be discovered and, because its output and effect are relatively easy to measure, it has been a popular topic of study ever since.

Under the higher regulation of the brain and the pituitary, the thyroid produces hormones that are intimately involved in growth. Tadpoles given thyroxine, a form of the hormone, metamorphose into grown frogs much more quickly than do normal frogs—although they develop into dwarf frogs. A species of gill-breathing Mexican animal, the axolotls, become land-living salamanders when subjected to the same treatment.

Possibly more important is the effect of the thyroid hormones on brain development. If the thyroid gland of new rats is removed, the rats grow up cretins, with considerable less branching of the brain than normal individuals. Similar effects also occur in humans: lowered thyroid activity in infancy leads to cretinism—low metabolic rate, mental deficiency, and dwarfism. A similar effect in adults goes under the name of myxodoema. Since cretinism in the young is caused by too little of this hormone in the organism, the disorder can be cured by administering it by injection.

Cretinism and related symptoms are due to hypothyroidism. The reverse, hyperthyroidism, an overabundant supply of the hormone, produces general overactivity and protruding eyes.

There are two main methods to follow the workings of the thyroid gland. The first one depends on the fact that the major part of the gland's activities involves the incorporation of iodine into an amino acid product. Therefore if radio-active iodine is supplied to the organism, its progress through the thyroid gland and subsequently along the target organs can be followed. The other technique depends on a correlation between thyroid production and what came to be called the basic metabolic rate (BMR). Essentially this indicates the efficiency of the organism at rest in converting its food intake into energy by measuring the amount of oxygen consumed, or carbon dioxide given off or, more directly, the heat output of the organism in terms of calories. The BMR measurements depend on a number of assumptions, some valid and some less so, concerning the efficiency of food utilization, a rate which varies greatly between individuals, and for different types of food. Despite its inaccuracies it does give a reasonable indication of the total metabolic speed of the individual. In this way it was found that over- or under-activity of the thyroid gland is reflected by a divergence of BMR of up to 20 per cent from the normal, and this result could also be maintained in a cell-free system.

Our main interest must belong to the ways in which the thyroid hormone carries out this speed-regulating effect. We have seen (see page 189) that in the respiratory chain the uncoupling action of some chemicals, notably dinitrophenol, can achieve the uncoupling of oxidation and phosphorylation and speed up oxidation. The action of the thyroid hormones is similar; it has been argued that they are therefore a type of uncoupling agent for the respiratory chain. To an extent, this explanation appears to be right. It seems that on the molecular level, thyroid hormones can indeed attach themselves to the cell mitochondria and achieve a degree of uncoupling; the results are seen in a rise of body temperature and thinness, as there is little ATP to synthesize fats and proteins. The increased rate of burning produces more end-products that have to be removed; thus more nitrogen appears in the urine. Lack of thyroid hormones would have the reverse effect, based on strong coupling in the respiratory chain and therefore an abundance of ATP.

There is evidence that thyroid hormones probably also act at a far more fundamental level, through the genes. This regulation is envisaged as going through two mechanisms; in one, the hormone activates the enzymes involved in the copying of the DNA strand by the messenger RNA, so that presumably more will be produced. The other mechanism is supposed to be the production of a more stable RNA, thus producing more long-lasting templates for protein synthesis. Both these effects would come under the general heading of coarse controls, since they depend on the subsequent synthesis of protein molecules.

Essentially, our views on hormone regulation change not only with the increase in the sum of available knowledge, but also with the general scientific ideology of the times. It can be argued that explanations have a tendency to be made in terms of the models current at a given time; at present the ideas of control through gene regulation are in favour. It is possible that the thyroid hormones, and also others such as aldosterone, act as gene repressors on the bacterial model we have looked at previously.

It is an observed fact that there are also direct effects on enzymes; thyroid hormones can dissociate—and thereby render inactive—glutamic dehydrogenase, an enzyme which converts glutamic acid into ammonia and alpha-ketoglutaric acid. Control mechanisms of this type probably function as fine regulators.

Thus it seems that hormones, at the biochemical level, can act in a number of ways connected, no doubt, with the interdependence of cells. The as yet obscurely-understood laws of cell contact must also have some influence on hormonal action, and indeed we shall only be able to get to the root of this problem by not considering cells in isolation but as parts of given tissues, or indeed the whole organism complete with its complex interactions.

Concentration on any single explanation has its obvious dangers. We know, for instance, that the drug puromycin inhibits protein synthesis. Thus, if puromycin blocks the effect of a hormone, it would be reasonable to argue that this effect was mediated through proteins. Another drug, actinomycin D, can stop the synthesis of RNA, and thus the copying of the DNA strand. If, therefore, the effect of a specific hormone is blocked by both these drugs, there is a superficial reason to argue that gene action must be involved. Not really. For the relationship is by no means so simple. We can no more say that these two drugs have only those effects we know about and no others than assume that the hormones act in only one way. It may well be that secondary effects have been at work and what one is observing is no more than a summation of diverse small effects.

What is unavoidable is that an explanation must be thought of in terms of a small molecule having large effects, directly or indirectly, on a collection of very large molecules which themselves are part of even more complex and interesting systems. It seems reasonable, therefore, to assume that whatever molecular control mechanisms there are, some spatial contact or effect is involved that will change configuration and thereby the environment in which certain reactions become more or less probable. The short answer to the exact problems of hormone control is that we have the ideas for an outline of an explanation but are a very great distance from a provable one.

NERVES, MUSCLE, AND BRAIN

The regulating functions of nerves and muscles have been well appreciated, if not understood, for a long time, for these two agencies are only too clearly involved in the most obvious human activities, and indubitably action takes place as the end result. Action is most obviously seen in muscle action, and nerve participation is its moderator.

In this area of the subject, as in most others, it is possible to begin with correct anatomical knowledge and progress through a better comprehension of the underlying physiology, until we arrive at the realization that chemical movement underlies and determines how muscle and nerve fulfil their task. How it does so is no easy matter to discover, for experimentally it is necessary to operate on very small pieces of matter indeed, so that unfolding understanding had to await the development of such techniques as the micro-electrode to measure minute currents, chromatography to describe the existence and distribution of chemicals, and all the other paraphernalia of large-scale, expensive modern research.

Even so, large areas of uncertainty still remain, and it is hardly surprising that this should be so. After all, the average human possesses thousands of nerves and muscles; not, alas, strung out in one straight line but interconnected to each other and to innumerable other tissue in the most fascinating and complex arrays, all working full time and full out on their respective tasks; and the brain presents complexities of such orders of magnitude that we still have only the haziest ideas on how it really functions. Analogies of the action of the brain tell us more about the current state of the material culture of those who propose the similes than about the thing to be explained. What is the brain: a system of complex mechanical levers, an array of telegraph machines, a giant computer, a series of electrical discharges, a memory of past RNAs? An explanation would not only be pleasant but also reassuring, because the brain and its accompanying nerve system is not only an organ but an idea of identity, a measure of our being. But about the brain we know very little, although what we know is highly interesting.

Starting from the easy end, some sort of nervous system is not the privilege of man alone, or indeed of the higher animals. The beginnings of such a system we can see even in the very lowest forms of organism—hydras or flatworms, for instance. A good general rule is that if the organism, in order to survive, has to make some response towards its environment which is more than the purely chemical or mechanical variety shown by plants, there must be some mechanism to register the changes in environment, some decoding mechanism to translate the effect of these changes into a series of commands applicable to the organism, and finally some means whereby a regulatory mechanism can be put into action. If, therefore, a simile is needed, the brain and nervous function, at least in the broadest outline, is similar to nothing more than those rather sophisticated factory control systems which go under the name of in-line or black-box control. Here, as in the sensory functions of the organism, there is first of all a transducer at a sensitive point that translates some environmental change such as pressure, temperature, or light into some easily transportable code. In the factory this is usually electric impulse or hydraulic pressure, in the organism almost without exception electric impulse. The message then travels to a control centre, which compares its value to another pre-set or memorized measure. Any deviation from this value calls forth correcting action. Such a system must be accurate, or something very tragic might happen. It must also be fast. If the pedestrian does not jump out of the way of the oncoming car in time, or the pressure regulator does not alter the valve setting, a quite irrevocable event will take place. In the living organism, receptors (the eyes, ears, touch) are activated by outside stimuli. The message is translated into a pattern of nerve impulses, basically an electrical phenomenon, and conveyed either to the brain direct, or to some less powerful clearing-house. Here the appropriate effect is put in motion and the tissue that has to respond is activated. The

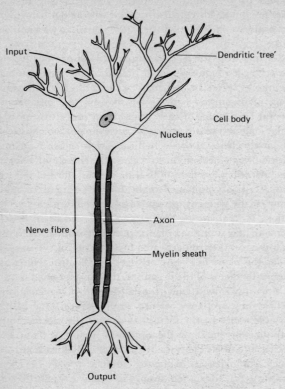

Figure 38   A Nerve Cell (much simplified)
Endings form 'synapses' (contacts) with other nerve cells or with effector tissues, e.g. muscles.

commands to the activated tissue also proceed along the nervous system, and in addition the interconnections of the system allow a general integration to take place. In other words, the specific spectrum of ancillary messages is also broadcast: for instance, the appropriate endocrine glands have to be brought into the picture.

The unit of the nervous transmission is the nerve cell. Obviously, nerve cells have taken elements of the common variety of cells, but their special functions have called for a great deal of adaptation. Thus, the tree-like dendritic region is adapted to the receipt of impulses which travel down the line from the environment to the axone portion. The message, having travelled through the body of the nerve cell, is adapted for the transmission of the impulse to the next functional unit, be it another nerve cell, a muscle, a gland, and so forth. There is a great deal of physiological interest concerning nerves and nerve transmission: for example, viewed in cross section the axone is rather like a telephone cable, with a number of individual nerve fibres enclosed in an insulator; in this case we have myelin, rather than the cable's polyvinyl coating. There are a number of such features: the way the axone divides into a number of strands at the end so as to increase the area of contact with the subsequent unit, the actual construction and function of the joint between nerves, called the synapse, the way nerves are linked, as in a relay system, to ensure that the message reaches an outlying tissue strongly. All these considerations are, unfortunately, outside the realms of biochemistry and belong to a variety of topics, from physiology to anatomy. Some of our ideas about them also effectively belong to the sphere of imagination.

But there is sufficient interest left if we consider only the biochemical aspects of nerve behaviour. Even here, we have to leave some of the most fascinating aspects of this problem, mainly because at present they are little understood. For example, what is the effect of the interaction between several nerve units, the neurones? If nerves can act as systems of relays, it is obvious that further nerves may well become involved in the transmission of the message. What then is the effect of the interconnections? And specifically, the brain being the largest assembly of interconnecting nerve tissue, can we talk about brain function at all in terms of the single nerve, or neurone? Should we not rather consider the brain as a whole, working in a specific environment?

Before we consider some of these ideas, let us look at a relatively simple case: a message being transmitted by one nerve. We can say experimentally that the message is electrical; more correctly, if we connect electrodes to specific points along the nerve we can follow the passage of a voltage variation. The difference in potential between the inside and the outside of the nerve adds up to about 80 millivolts.

The source of the difference in potential is a mechanism analogous to that operating an electric battery. In the medium outside the nerve cell we have a given concentration of salt, that is, of sodium and chloride ions. Inside the nerve, there is about a tenth as much salt, but there is an appreciable presence of the potassium salt of some organic acids. The reason that these substances do not simply become mixed is that the nerve membrane does not effectively allow the entry of sodium or the exit of potassium. In other words, it acts as a specific barrier, or in more picturesque language, it can recognize sodium and potassium. The recognition depends on the size of these ions. It is interesting to note, however, that such specific membranes are not the unique property of biological systems; recently, for example, a plastic sheet has been developed which allows the ingress of oxygen but not that of water, so that it can theoretically be used under water by divers.

In the nerve we have a situation in which there is a net negative charge on the inside of the membrane because potassium is more negative than sodium. By the simple application of electrostatics there must be a net positive charge outside the membrane. If, for whatever reason, an electrical impulse is applied to the nerve, the system will be upset and the accurate equilibrium between charges on the opposite sides of the membrane will be destroyed. This so-called depolarization will propagate itself along both the inside and outside of the nerve, since the change of sign of a charge has the inevitable consequence that surrounding charges also have to change. Thus a series of charge reversals will be propagated along the nerve fibre: a small electric current will pass.

The disturbance of polarity also disturbs the specificity of the nerve membrane. When depolarization first occurs, sodium ions rush in, lowering the difference in potential between inside and out. Indeed, they overshoot zero; what previously was a potential difference of −80 millivolts becomes +30 millivolts. But this very process defeats its own completion; as the concentration of sodium ions on the inside of the membrane rises, the accumulating charge will prevent the further ingress of the similarly positive sodium ions. At the same time, potassium ions are being pumped back into the inside so that conditions of equilibrium eventually re-establish themselves. Although this process may sound complicated, in actual fact it takes less than one microsecond before the impulse passes on to the next section of the nerve.

Propagation of nerve impulses depends, therefore, on the passage along the fibre of an electrical disturbance, a depolarization, which can be seen as a changing pattern of sodium and

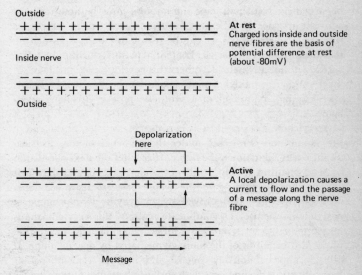

Figure 39 The Nerve in Action

potassium ions. As the basis of this function we must have, therefore, a regulating mechanism, and when we look for chemical agents, we come across our good acquaintance, ATP. How ATP comes into this system is still a matter for experiment and controversy. It is undeniable, however, that ATP is implicated in the enzymes that probably account for the changes in shape of the cell membrane which regulate the entry and exit of the two metallic ions involved. How it does so is not clear, but if we poison a nerve with cyanide, an injection of ATP will restore its function, a result that seems to indicate that the enzyme-ATP complex is intimately involved in these processes.

The next obvious answer biochemistry must give us concerns the problem of how all these processes start. In order that a portion of the nerve membrane may be depolarized, a charge must have reached it. Where from? Here we must consider the synapses, junctions between nerves and between nerves and other tissues. In anatomical fact, where the axone ends and the subsequent nerve begins, there is a small gap; the impulse must be able to jump across it in order that the next tissue may carry on. In some cases this gap is so small that we can be reasonably happy postulating that the difference in electrical potential permits the impulse to get across without any help. In others this cannot be so, and here we meet another of those key materials in biochemical work. This is acetylcholine, a product of a synthetic process involving coenzyme A.

Acetylcholine is stored in little packages, vesicles, at the end of the axone, and is released under the influence of the electrical impulse travelling down the nerve. It reaches the next nerve and, by a mechanism which is still not understood, causes depolarization, thus starting again the whole process. Its work done, acetylcholine itself is destroyed by an enzyme by the name of acetylcholinesterase. To complicate matters, this very enzyme is also supposed to act as a receptor for acetylcholine and become the physical transmitter of the message mediated by acetylcholine. If all this appears unlikely, one should perhaps remember that basically one is dealing here with surface phenomena; the absence or presence of certain chemicals distorts the reaction surface first one way, then another, so that different reactions can take place between the same participants in a theoretically identical environment.

There is a fair amount of experimental evidence that these ideas are along the right lines even if their details are wrong. For example, it is possible to use drugs to inhibit the action of acetylcholinesterase. The result is that acetylcholine cannot pass from one nerve to the other, paralysis results, and if the nerves happen to be ones controlling a vital organ, the end result is death. Other drugs, such as nicotine and curare, have similar effects.

We can turn this problem around and consider the possibility not of nerve-activating substances, but nerve-inhibiting materials. It turns out that such chemicals exist, one of them being gamma-aminobutyric acid, GABA. Its action appears to be to stop the excitation of nerve no. 2 as a result of excitation of nerve no. 1 (Figure 40), presumably through a mechanism analogous to that

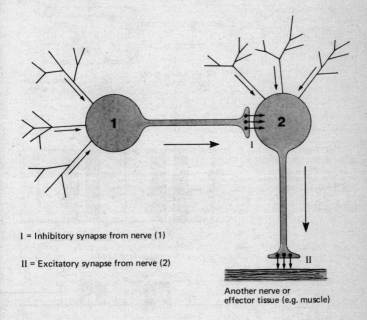

I = Inhibitory synapse from nerve (1)

II = Excitatory synapse from nerve (2)

Another nerve or
effector tissue (e.g. muscle)

(Arrows indicate direction of nerve impulse transmission)

Figure 40 Excitation and Inhibition

Nerve (I) synapse releases a substance, gamma-amino butyric acid, which inhibits the action of nerve (2). The chemical transmitter for excitation, released by nerve (2) is usually acetylcholine. Thus, the release of specific chemicals mediates transmission of inhibitory and excitatory nerve impulses.

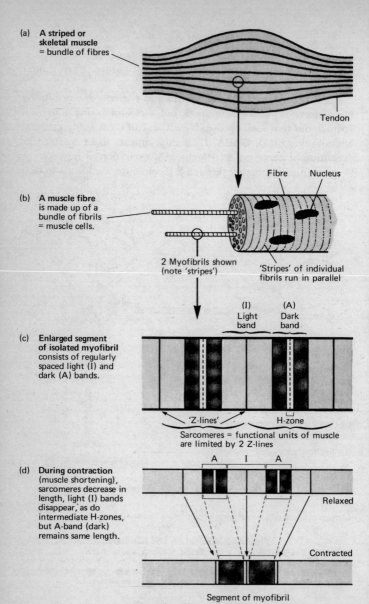

(a) **A striped or skeletal muscle** = bundle of fibres

Tendon

(b) **A muscle fibre** is made up of a bundle of fibrils = muscle cells.

Fibre Nucleus

2 Myofibrils shown (note 'stripes')

'Stripes' of individual fibrils run in parallel

(c) **Enlarged segment of isolated myofibril** consists of regularly spaced light (I) and dark (A) bands.

(I) Light band (A) Dark band

'Z-lines' H-zone

Sarcomeres = functional units of muscle are limited by 2 Z-lines

(d) **During contraction** (muscle shortening), sarcomeres decrease in length, light (I) bands disappear, as do intermediate H-zones, but A-band (dark) remains same length.

A I A

Relaxed

Contracted

Segment of myofibril

Figure 41 Muscle in Action

of acetylcholine. GABA is formed by a reasonably simple chemical reaction, decarboxylation, from a common amino acid. However, this reaction requires the presence of vitamin B_6, and if for some unknown reason this does not exist in sufficient quantities in the organism, the result is overexcitation. It has been found, for example, that newborn babies who for one reason or another have insufficient vitamin B_6 suffer from epileptic fits directly traceable to this deficiency.

MUSCLE ACTION

Muscles consist of bundles of fibres about 0.01 cm in diameter, made up of smaller fibres, called fibrils, measuring about 0.0001 cm across. Fibrils, or myofibrils, are the actual unit of muscle (Figure 41, parts *a* and *b*). Muscle cells have become highly specialized, so much so that when we place them under the microscope it is not even possible to readily distinguish the usual cell elements such as the nucleus. Microscopic work, later supplemented by X-rays, did, however, provide the basis of our knowledge of muscle action.

If we look at muscle fibres in a polarizing microscope—where the vibrations of the light rays are restricted to one plane only—we see dark bands, called A-bands, which alternate with light I-bands. Looking closer, we see that each I-band filament is surrounded by three thick A-bands, and conversely, each A-band has six light I-bands for next-door neighbours. The I-bands are connected to the A-bands by tiny bridges, and in the middle of the A-bands there is a lighter zone, called the H-zone.

When the muscle contracts, the A-bands stay put, but the I-bands move towards the A-bands so that the previously light H-zone becomes dark. In normal contraction the muscle can shorten to about 60 per cent of its original length. We can describe this movement in slightly different terms. It can be shown that each myofibril consists of thousands of small regions called sarcomeres, and it is these regions that provide the necessary movement for contraction. (See Figure 41, parts *c* and *d*.)

The explanation we have described is known as the sliding filament theory, based mostly on biophysical work and having a great deal in common with geophysicists' explanations of some types of earth movements.

Seventy per cent of muscle protein is composed of myosin (about 50 per cent) and actin (20 per cent), and our biochemical explanation of muscle contraction revolves around these two materials. Actin exists in muscle in two forms: fibrous and globular. We can also assume that the globular form is a protein monomer, but that the fibrous version consists of polymerized protein molecules. We may now tie together biochemistry and physiology. The I-bands seen under the microscope consist of fibrous actin while the A-bands are myosin. During contraction, globular actin polymerizes to fibrous actin accompanied by a breakdown of ATP, also present in the system. The molecular biology of muscle contraction, based mainly on the work of Huxley and Hanson, is known as the 'sliding filament' theory. This is shown diagrammatically in Figure 42.

Thus the energy for muscle contraction comes from systems involving ATP that in turn can be produced by a precursor, creatine phosphate, a phosphate-containing amino acid derivative. The production of ATP, as we have noted, derives from cell respiration; thus it is this species that provides the link between the energy-using and energy-making reactions of the body. If energy is urgently required during muscle exertion, ATP is produced directly from creatine phosphate rather than from respiration, which would take too long.

There are several pieces of experimental evidence accumulated over the years that form the basis of the biochemical theory of muscle action. In the 1940s Szent-Gyorgyi and Straub in Hungary noticed that actin and myosin extracted from muscle can be made to contract in a test tube if supplied with ATP, but neither protein worked on its own. Some years previously Russian workers had shown that myosin can break down ATP by means of an enzyme called ATP-ase.

Thus the overall view of muscle contraction asserts that in the resting state ATP is bound to myosin and ADP to actin. During contraction the myosin–ATP complex breaks down, helped along by ATP-ase; globular actin polymerizes, and the chemical bonds

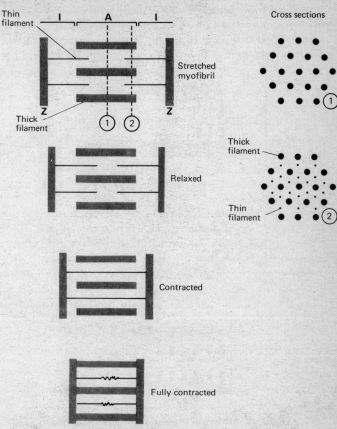

Figure 42 'Sliding-Filament' Theory of Muscle Contraction

In the electron microscope, a sarcomere is seen to consist of thick and thin filaments which account for the light (I) and dark (A) bands.

During contraction, these filaments do *not* change in length, but the thin filaments 'slide' between the thick ones, increasing their overlap, until only dark bands are seen. The thick filaments are mostly molecules of myosin and the thin ones, actin, the two main fibrous proteins of muscle. Movement of the filaments past each other is energized by the breakdown of ATP by myosin, and thus myosin functions as an enzyme ATP-ase. ATP breakdown causes molecular 'hooks' between the actin and myosin filaments to break and reform, pulling the actin filaments by a ratchet-like mechanism.

between the two proteins are differentially broken and made. The whole sequence is triggered off by calcium ions released by acetylcholine supplied by nerves, or by direct nerve action on the muscle membrane.

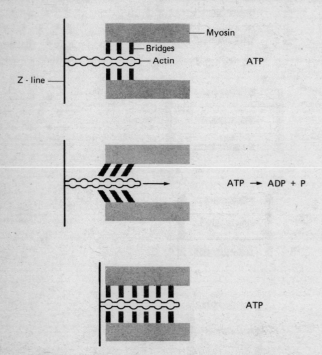

Figure 43 Molecular Biology of Muscle Contraction
Detailed mechanism of muscle contraction is based on the 'making and breaking' of bridges formed between thick myosin filaments and thin actin filaments, the process being 'energized' by ATP breakdown. These bridges are detected in electron micrographs.

The whole process appears to be initiated by free calcium ions released by electrical or chemical (acetylcholine) stimulation of muscle fibre. Later, calcium ions are reabsorbed into a bound state during relaxation. This is eventually a ratchet-like sliding of filaments.

THE BRAIN

All that has been said about the complexities of biochemical systems is a thousand-fold true of the brain, an organ that by its very nature excites not only scientific curiosity but also mystical awe.

Currently biochemical research on the brain, despite the intricacies of its experimental techniques, is only at its beginning. We have just reached the stage where increasing knowledge underlines the vastness of our ignorance. Conversely, biochemical research on the brain offers to the young hopeful a field in which the really significant discoveries are still to be made. There are very interesting areas of research, some of which we shall discuss briefly.

The brain is a highly developed organ, possessing a range of biochemical mechanisms and physiological functions; clearly there must be intimate relations between the two. There ought to be, for instance, some correlation between mental illness and biochemical cause or effect, and a great deal of research work is spent on finding biochemical manifestations of mental diseases such as schizophrenia. Still, the correlations are not entirely satisfactory and are usually the subject of a great deal of dispute. It has been found that when certain drugs are injected in man or animals they can produce behaviour comparable to that shown by schizophrenics, and therefore research is being carried out to find unusual material in the blood or urine of those suffering from this disease. In particular, since drugs resembling neurohormones, such as adrenalin and noradrenalin, can cause odd behaviour, it would be expected that they, or their successors would turn up in the patients' urine. On the reverse side, it has also been found that some drugs that alleviate the symptoms of schizophrenia may well be acting on the metabolism of particular types of compounds such as the amines.

Unfortunately there is as yet no complete agreement on the disease itself. For example, we are not completely clear which are primary and which are secondary symptoms, or what exactly is cause and what effect.

Attempts to find a biochemical basis for mental disorders are proceeding in a number of areas. In all instances the underlying assumption is that mental disease is but an expression of some metabolic, possibly localized disorder affecting some part or function of the nervous system. That this assumption is roughly correct is shown by the effects of drugs on personality and behaviour. Thus it seems likely that if one could relate at least the gross outlines of some metabolic pathway to brain function, one could proceed a long way in providing therapeutic methods even without accurate knowledge of the pathway in question.

The field, however, is still open to wide-ranging, imaginative but as yet, factually unsupported speculation. Linus Pauling, for example, suggested that mental illness may be caused by the lack of certain micronutrients in the diet—which presumably would be present in so small or so transitory quantities that they would not have been recognized. This idea stipulates that mental illnesses are deficiency diseases in the same way that, for example, scurvy is. Sufficient evidence relating micronutrient deficiency and mental illness has not been found, and the verdict must be not proven, although it is known that the classical deficiency diseases may have not only gross physical, but what may be described as psychological effects.

Another theory considers that a 'twisted protein' is the main reason for mental illness. In this case also, we have experience of what a 'malformed' protein can do—in sickle-cell anaemia, as we have seen, an alteration of just one amino acid in the protein sequence can result in the pathological condition. If this theory is correct, it is possible for the rogue protein to cause damage directly or through the disruption of a metabolic pathway leading to an accumulation or deficit of some intermediary.

At the same time a great deal of research is also taking place to characterize mental illnesses more precisely. Thus one would not be greatly exaggerating to suggest that the final solution of the problem of mental illness will be the outcome of a race between biochemists and clinical psychologists, each following up the problem from their own base. From the strictly logical point of view, there is reason to expect that the methodology of biochemistry may well achieve results earlier. This will involve the complete consideration of the biochemistry of early development, especially of the brain, to determine the factors relevant to its later function.

THE MEMORY

One aspect of brain function that has received considerable publicity is the mechanism of memory. Here again we are in an area where scientific regard is joined by emotional drives, and the results are often very confused.

Some years ago, workers in the United States claimed that not only could flatworms learn by being conditioned but that wisdom thus gained could be passed on by biochemical means. The worms were conditioned to patterns of behaviour, for example, to find their way around a maze or to show a given reflex, such as contracting when a light was flashed. This learning process required an observed amount of repetition. The wise worms were then killed off, chopped up, and fed to another set of untaught, or naïve worms. Lo and behold, the naïve creatures set about their learning with greatly enhanced facility and became wise in a much shorter time than could reasonably be expected.

The rationale behind these findings was that the naïve worms ate a somehow modified RNA in their unconventional diet. The wise RNA was in some concrete, physical way modified and, when ingested into the system, carried the patterns of knowledge with it. In support of this argument the observation was cited that when the destructive enzyme RNA-ase was present in the water the worms occupied, the biochemical learning process did not take place, suggesting that the wise RNAs were destroyed.

The worm experiment caused something of a furor. The correctness of the observations, let alone their interpretations, were severely questioned or flatly disbelieved. Yet arguments do not make a scientific truth, and the plain fact is that nobody can prove or disprove the case of a learning process through the physical modification of some measurable quantity.

The worm runners—whose first results were published in the coyly named 'Worm Runners Digest'—then went on to rats. Being somewhat more intelligent than earthworms, rats can be taught a number of tricks; in the language of the art, they can be conditioned in various learning situations. For example, they can be taught to press levers, or to run mazes, or to avoid electric shock. The rats were taught one or more of these tricks and were then killed and their brains extracted and mashed up. The RNAs or proteins were extracted and fed to untrained rats, who thereupon proved to be far more intelligent and teachable than could be expected.

The biochemical fracas thereupon started again, augmented this time by a row over the nature of the agent that transferred the learning, RNA or protein. And, as usual, every step from observation to conclusion was hotly contested.

Undoubtedly, there is some factual residue at the basis of these experiments. Learning, at least long-term memory, must have some physical basis, and therefore involve the concrete modification of some entity. This general argument is supported by the observation that drugs can inhibit memory, although whether this is a primary effect or only a secondary consequence of the brutal churning-up of the whole system is not known. It is reasonably probable that the physical entities undergoing changes as a result of learning are either RNAs or proteins, since if the synthesis of either of these is inhibited, the learning process stops. Taking this idea a stage further, one can argue that RNAs produced by DNAs in the neurones of specific nerve circuits are somehow modified to take an imprint of past events. This idea would fit in with the compartmentalization of the brain; thus there would be specific areas for the storage and subsequent manipulation of auditory, visual, olfactory, and other sensory events. All this, of course, is pure speculation, although we do know that increased activity in one area of the brain is accompanied by increased and localized RNA production. The question is, which is the consequence of which? Once increased RNA synthesis can be correlated with learning, we can extend the argument to suggest modifications of the synapses, and thus elaborate a mechanism depending on an increased information-carrying capacity of nerve fibres, possibly in terms of a greater number of impulses.

Before becoming too bemused by all this speculation, let us remember that a perfectly good biochemical memory process exists in the organism outside the brain, in the shape of the immune system. We recall that once the organism has been alerted by the presence of an antigen, specific antibodies will be produced even when the antigen had been unknown to the organism. Once the production of these antibodies has been learned, they will always be produced in response to that antigen, even when similar events are separated by the passage of many years.

It can therefore be argued that, in a sense, the capacity for learning may be a characteristic of certain biochemical systems which is called forth by specific situations. What we normally understand by learning may be no more than the manifestation of this characteristic in a highly complex environment supplied by the intricate nerve interactions of the brain. But at present we still have to wait for the answers to these questions.

CHAPTER 8

perspectives

A collection of facts does not make a science, any more than a builder's stock constitutes a home. Not only have materials to be sorted into logical sequences but some underlying plan has to illuminate the whole process. The end result depends not only on the excellence of the parts and the care with which they have been put together, but even more on what is no more than a dream in some creator's eye.

One should beware of taking analogies too far, but all enquiries into the processes of scientific work agree that it is a great oversimplification to concentrate on factual, logical sequences to the exclusion of all other considerations. Behind the facts there has to be an idea, behind the idea some perception of order, a creative process similar to that experienced by the great artist or writer.

Science is a many-splendoured activity: observation, empirical knowledge, hypothesis, theory, verification, and development all proceed simultaneously, although during the history of a discipline now one, now another of these activities demands maximum interest. In the beginnings of a discipline the observational, fact-finding stage is the most important. Knowledge is so sparse that large amounts of data are required to supply a framework for thought, to delineate the boundaries of the discipline, and to allow the emergence of concepts that will enable the scientist to discuss his ideas in meaningful language. Waxing and waning in importance, the fact-finding stage is a cornerstone of any discipline. It is never forgotten; it is constantly pursued.

Our best example comes perhaps from physics, where since the time of the ancient Egyptians facts were noted and listed, although theory developed and withered with the changing attitudes of contemporary civilizations. Yet even now, when physical theory reaches levels of mathematical complexity understandable only to a small minority, the search for more facts goes on, accompanied by the need for ever more complex and expensive apparatus.

In the same way, biochemistry and the organic chemistry that preceded it contain a large element of fact-finding. From this point of view, biochemistry is still a cheap science in which costs are counted in thousands rather than in millions of pounds—an important factor both for its practitioners and for society which must finance, utilize, and accept its results.

During the years when data are newly discovered and their implications slowly digested, the excellence of theories is of less importance. Contemporaries will argue, and since they are interested in and emotionally involved with their subject, will argue passionately. But looking back we can observe the history of the subject and ruefully note that in their particular contexts theories which often turned out to be incorrect frequently produced as much of a hard core of fact as did the correct ones.

It used to be fashionable to decry the ideas of the alchemists who spread their own blend of torrid logic over the intellectual life of the civilized world. Yet, shorn of all mystical pretentiousness, these protochemists contributed a great deal to what we now regard as chemistry, chemical technology, and chemotherapy. Whether a liquid boils because the spirits of the vapour try to escape, or because the molecular cohesion decreases, is after all a luxury of thought. What matters in the first place is that we should be able to boil the liquid safely and even distil the various fractions as and when we desire.

Thus the human mind first wants to know what happens. But even when facts are at its disposal, it is unsatisfied; the second set of questions, well known to any parent with small children, is the magic incantation: why? We in the West have opted for a general view of the world that takes as its central dogma that every event must have a cause. Causality may be hidden by our ignorance, but it must be present. Perfectly good systems of thought have been constructed even without this central premise, but having accepted it, we must fit our facts into logical, causal patterns.

A logical sequence that postulates a reason behind the event is called a hypothesis, or the construction of a model. A model must be amenable to scientific investigation, it must have predictive powers that may be checked by successive experiments. But although the consequence of the model is a series of logical actions that will modify the hypothesis and lead it nearer to reality, it must be argued that the genesis of the hypothesis, expressed in a model, owes a great deal both to scientific experience and to insight that can only be described in terms of intuition. The essence of a model is that it provides the pattern for existing facts on which further research can be built, but it also has a further consequence. It conditions the thinking of those who accept it, and thus its influence may stretch far beyond the immediate application. Bohr's model of the atom was framed to answer problems thrown out by spectroscopic measurements; its implications reached all the way to the philosophical bases of determinism. In the same way, the current DNA model conditions biochemical and biological thought on a wide front.

A model can be a very simple one: atoms are little hard balls that spring apart on collision. Or it can be a vast mathematical edifice that needs a computer to solve it. The model tells us that certain systems work in a given way under specified conditions; thus, knowing the way enzymes act, we are able to predict their effect on certain materials at a given temperature in a specific solution.

But however complicated and scientific a model, we must beware of attributing more to it than we have designed into it. The best thought-out computer programme will not answer questions we have not expected to ask; all it can do is to reshuffle information and present it in a different form. Biochemistry is currently in the state of model-making. DNA can be described as a double helix, protein synthesis may be made sense of if we postulate the sequence of gene, RNA, ribosome, and so forth. We must bear in mind that models are aids to description, logic, and the devising of further experiments. Time will confirm them or illuminate areas needing revision. Models may be, but are not necessarily exact descriptions of reality. In physics the mathematical model of gravitation developed by Newton stood unchallenged for three hundred years. Ultimately it proved to contain truth but not all truth. In biochemistry, the era of model-making has only just begun: it will be unnatural if all our explanations of today will survive intact the test of time.

Beyond the explanations derived from models and confirmed by experiments comes a stage in scientific thought that may be expressed as the general ideology of the discipline. This is uneven ground, for many will argue that the role of the scientist is to confine himself to his subject and, by patient gathering and interpreting of results, to act out his role in enlarging the sum total of human knowledge. This, one must argue, is a restrictive point of view, proven obsolescent by the very environment science helps to create.

For the scientist, and especially for the biochemist, a wider role must be faced. He must submit to the requirements demanded of the scientist in his professional role, but he must also realize his rights and duties in a far wider social environment. As a professional scientist he must obey the rules obligatory for all practitioners of repute. He must, for example, be scrupulous in the representation of his results and the inferences he draws from them. He must not advertise himself by making extravagant assertions, claim originality where there is none, or appear to disregard the contributions of his peers. Unfortunately, he will also be expected to use the formalized and often convoluted jargon of his discipline and to pretend a logical sequence of theory-experiment-improved theory that often does not exist. All these hierarchical gestures that confuse or amuse the outsider do have a serious purpose: to ensure, if not peer approval, at least acceptance of the general scientific community. Acceptance does not imply agreement, but it does afford a common language and a chance to carry on meaningful discussions. Science itself is strongly evolutionary, with occasional revolutions. The organization of science is arguably conservative. But even if it were as modish as yesterday's art vogue, there would still be no place for the immature iconoclast, for the simple reason that it takes time to learn which icons to smash. The scientist, in order to communicate with his fellows, must show responsibility through knowledge to the scientific community.

Relationships to society at large present an equally great measure of responsibility. In the past there were few scientists, their effect was minimal and, what is perhaps most important but often forgotten, the prosecution of science was not economically significant. The actions of scientists mattered little. But today, under influences directly traceable to science itself, the situation has completely changed. In the 1920s Rutherford could truthfully say that he found no reason to think that the splitting of the atom would ever have uses. No scientist, in whatever area he is working, could be absolutely sure of the truth of such a statement today.

Biochemistry has a particularly sensitive position among the sciences, and responsibilities that have hardly been fathomed, let alone faced. The intricacy of its position arises both from the complexity of factors ruling the conduct of any living matter and from the obvious consideration that anything biochemistry tells us concerns our very existence. No wonder that aspects of the life sciences have always been influenced by emotional, mystical, and even political considerations. The recent Lysenko controversy in the allied field of genetics should be a reminder that extraneous influences are by no means mere past historical accidents. And indeed it is proper that this should be so. If science can cure cancer, arrest disease, and prevent death, it must also declare itself about the care of the living, an ageing population, or its potential interference with our genetic make-up.

It is undeniable that the extraordinarily rapid progress of science has brought about a reaction against it, especially among those who should be more than ever cognizant of its beneficial power. The signs of this revulsion are all around us, even if only as yet to a small extent. A significant statistic in the United Kingdom shows that whereas in the past few years the proportion of A-level pupils in schools choosing science has fallen, in the last year even the absolute numbers have been decreasing. It is easy to argue that most intelligent young people are aware that their future chances are better studying subjects relating to human behaviour, such as sociology or economics. It is also said that these disciplines offer an intellectually less demanding path to the prospective university graduate. Yet the idealism of the young, however ineptly expressed, should not be undervalued. Revulsion from science is a perfectly natural gesture from those who, however dimly, are conscious of the potentially destructive power of science and who are therefore groping towards means of studying and controlling this power.

Neither is it a meaningful argument that biochemistry, in common with all the other sciences, suffers from a total confusion between the nature of science and technology in the minds of its spectators. Technological use or misuse is brought to the doorstep of science with a resultant harm to both. This dichotomy can be seen very clearly in the case of chemistry, physics, and engineering. Chemistry will tell us the properties of the reactions of explosives. Physics will acquaint us with the properties of materials under a range of conditions. But rocket research, for example, is neither chemistry nor physics: it is engineering. Biochemistry may tell us about the fundamental problems of animal tissue. The application for specific purposes of an agent discovered to control cell metabolism is no longer biochemistry. The discovery may be used for the relief of cancer or it may conceivably be used for warlike purposes.

If developments in biochemistry are to mean anything, they will have to be translatable into technologies that at some remove may be applied to the individual. The mass manipulation of thought processes and the breeding of specialized clones of sub-humans are still in the scientifically untrained minds of the professional crystal-gazers, but moral and ethical problems that may arise with further developments in biochemistry are already with us. They should be faced before the time comes when judgement will have to be made on specific and urgent issues.

There are currently two fundamentally opposing points of view on the scientist's role in society. It is argued that the scientist must not be concerned with the eventual use of his discoveries once they are available to society at large. Conversely, it is argued the scientist has some divine right to inform his fellow citizens if their morality is slipping. It seems self-evident that in the long run neither of these oversimplified arguments can be tenable. Neither provides a true answer to the demands of a social environment that not only has to live with the products of its scientists but also is under an obligation to pay large sums of money to obtain them. Some compromise solution will have to be found, and it is perhaps already emerging through the various consultative mechanisms that both government and industry are employing to an increasing extent. Biochemistry, a relatively new science, untrammelled by the historical accretions of older disciplines, may have a role to play that will overshadow in importance even the strictly scientific aspects of the subject.

PAST AND FUTURE

Biochemistry grew out of the rapidly developing schools of organic chemistry that flourished mainly in Germany and France in the latter half of the nineteenth century. At this time the main emphasis was, as it had to be, on the isolation and analysis of substances obtained from the living organism, although studies of biochemical mechanisms can be thought of as starting with Pasteur and his work on fermentation.

Although the primary interest of chemists was in their powers of accurate description, the materials isolated did prove to be of great biological significance. With the accumulation of data, interest shifted towards finding out not only the chemical nature of biologically active materials but also their function. Thus arose the great schools of biochemistry, associated with names such as Fischer, Ehrlich, Wieland, and Warburg, among others. Once a school of world reputation is formed it acts as a forcing house for more first-rate scientists, thus spreading its influence and general interest in the subject. For political reasons the school may migrate: the revolutions of 1848 in France and Germany or the Nazi rule nearly a hundred years later provide good examples of political forces scattering scientists into happier environments. Butf the work of the school goes on and in turn generates the new non-conformists to press their own contributions.

The analytic and synthetic approach to biochemistry is a major force in the development of the discipline. A remarkable influence has been the influx of chemists trained in the handling of large molecules who turned their attention to the borderline between what a chemist calls a natural product and what a biochemist recognizes as a functionally interesting entity. Helped by the increasingly accurate, ingenious, and labour-saving tools at their disposal, natural-product chemists now follow in the footsteps of Wöhler, the synthesizer of urea. Synthetic organic chemistry is now associated with, among others, Todd in Cambridge on nucleotides, Woodward at Harvard on the synthesis of chlorophyll, and Khorana at Western Reserve on work that may eventually enable us to synthesize genes.

Our ability to analyse significant substances of the living organism, using chemical and biochemical tools, is being translated into the skill to synthesize them. This domain of work, still in its infancy, is one where some of the most important discoveries of biochemistry will undoubtedly take place in the years to come. Because our ability to synthesize will also become synonymous with the possibility of interference, it will automatically require much thought in considering the applications of discoveries. Already Sir Macfarlane Burnett, the Australian virologist of world reputation, has proposed that work on molecular biology should be stopped since the chances of virus mutations giving rise to species against which there will be no defence are becoming greater. Most biochemists would not agree with this view, although all concur with the need for strict controls; those who work in laboratories would, after all, be the first to suffer.

The history of insulin provides a good example of the analytical-synthetic method in operation. Discovered by a series of clinical trials on dogs, insulin became an accepted agent in the treatment of diabetes long before its constitution was known. The structure of the molecule was determined in Cambridge by Sanger, who announced his results in 1951 after years of laborious work. Finally, a group of mainland Chinese workers recently synthesized insulin in a form possessing all the physiological and therefore chemical characteristics of the natural product.

X-ray crystallography has been one of the most important aids in the discovery of the architecture of biologically important compounds. It was first used to a notable extent by Bernal in Cambridge around the thirties of this century, but before the development of computer methods to deal with its output, the technique posed formidable experimental and computational difficulties. It took nearly twenty years for the first complete analysis of a protein molecule to be achieved, by Kendrew on myoglobin, followed by Perutz on the larger haemoglobin. Apart from their scientific significance, these results had a number of other consequences. They provided agreement with Pauling's proposition that proteins are helices stiffened by hydrogen bonds, and in turn influenced our thinking about large biochemical entities in general. By demonstrating the power of the X-ray technique they also encouraged the researches that found their final expression in the Wilkins, Crick, Watson model of the DNA molecule. This in turn found an echo in the biochemical and genetic studies of Monod, leading to the current work of Khorana and others in attempting to synthesize the DNA molecule itself and thus gain more understanding of the mechanisms of genetic transfer of characteristics.

Computer techniques speeded up the interpretation of X-ray data, with a consequent increase in the number of new structural maps for proteins, mostly enzymes. Phillips and his group were the first to describe the details of the enzyme lysozyme, followed by other British and American workers. Currently the structures of the enzymes ribonuclease, chymotrypsin, carboxypeptidase, carbonic anhydrase, and papain, among others, have been elucidated.

This is only a beginning. A structural description is static, but in enzyme chemistry one is also interested in how catalysis is achieved. If an enzyme is in close association with a particular substrate when the act of catalysis takes place it should be possible to examine the disposition of chemical groups and compare them to their state when the enzyme is inactive. This can be done, provided both the pure enzyme and the enzyme-substrate complex can be brought to a crystalline state amenable to X-ray investigations. By evolving methods that show up variations over distances of one to two Ångstroms, it has been possible in some cases to build up a picture that correlates changing enzyme structure with its mode of behaviour. These data could then be compared with information from kinetic and other studies.

The original biochemical approach can thus be supplemented by essentially physical methods. The limitations of X-ray work are becoming evident: for example, its inability to help in situations where the overall picture must extend to an area more than 300 Å across and still be specific enough to show at least some individual atoms. Other physical methods are brought in: nuclear magnetic resonance spectroscopy, optical rotatory dispersion, the ability to measure minute currents and voltages, all of them measuring some property bearing on particular atoms in a molecule. These methods produce data in a complex form that has to be analysed and interpreted before its significance can be tied to structural or mechanistic features. At the same time, the use of these techniques reduces the differences between what is conventionally called biochemistry and biophysics.

It is perhaps fortunate that the application of these techniques to a great number of chemical problems is already routine. Here the borderline between chemistry and physics is becoming more and more arbitrary. The tendency for using more and more sophisticated hardware now reaching biochemistry has already become established in chemistry and has long ago overtaken physics. As the subject becomes more mature, measurements become more precise but also more complicated. Rewards may be greater, but there is also a danger that good ideas are lost among the paraphernalia of hardware. Even more importantly, as the apparatus required becomes more expensive, the limitations of research are defined not by the availability or relevance of problems but by the need for research funds to finance their prosecution.

The contributions of X-ray work to the elucidation of DNA structure are well known. RNAs present a more difficult problem because it is difficult to obtain them in a crystalline form, suitable for X-rays. Recently, however, crystals of a tRNA have been produced, which suggests that at least this particular specimen, that of a bacterial tRNA, has a cloverleaf structure.

Although it is very pleasant to know what particular

biochemical entities look like, structural determination is only part of a far greater challenge: the correlation of structure and function—what parts of a given substance carry out their tasks in particular situations. Once we have a good idea of both these aspects of behaviour, we shall be in a reasonable position to synthesize such molecules and even, should the occasion arise, improve upon the natural model. The synthesis of the DNA molecule is currently receiving a great deal of attention. At present laboratory experiments have to use natural products for templates, but there is no doubt that eventually we shall be able to dispense with these aids to produce genetic sequences at will.

All this is in the long term. Presently work is carried out on viruses and bacteria, and even the most sanguine biochemical optimist would agree that it is a long step between such organisms and the higher reaches of the animal kingdom. Nevertheless it is only a question of time before we shall have the ability to eradicate genetically caused disease and imperfections. We shall then also have the unenviable task of deciding which genetic make-up should be interfered with.

METABOLISM

The second aspect of biochemical studies centres on movement: what happens as foodstuffs are ingested and are turned into raw materials necessary for the organism, into work, and into waste products that have to be eliminated. Starting with Pasteur's work on fermentation, metabolic studies got into their stride in this century with the realization of just how important enzymes are in all cell metabolism. The idea of catalysed chemical reactions inevitably demands the consideration of the speed of reaction of energetics. This came in the thirties and forties of this century. Then came the realization of DNA structure and the coalescing of studies from a number of directions leading to a picture of the cell's control over its metabolic processes.

A further step along this road is to study not only how an individual cell controls its production of enzymes but how large arrays of cells do so and how such activities affect large-scale, measurable characteristics. An extension of this argument leads into the realms of interrelationships of the control of cell metabolism, cell differentiation, and neurobiology, the biochemistry of nerve and brain action.

Knowledge of metabolism has obvious implications for our understanding of the action of drugs and therefore our ability to design better and more powerful medicines. The development of drugs is still to a large extent an empirical exercise, with obvious and necessary waste as clinical trials screen out those not effective enough or with unacceptable side effects. As our knowledge of the specific action of drugs increases it will no doubt be possible to dispense with a great deal of our current overinsurance and start with possibilities having a far greater chance of ultimate success.

With developments in surgical techniques, the problems of graft rejection will undoubtedly increase. The solutions are tied to considerations of enzyme structure, metabolism, the nature of immune and auto-immune reactions, and to the ability of substances to recognize others. Because in both memory and immunological functions aspects of chemical recognition must be involved, it is not unlikely that results from researches on one of these problems will have implications for the other. As antibody structures and mechanisms are elucidated it will become possible to formulate agents that will allow the transplant of organs by decreasing the severity of the rejection mechanisms and at the same time allow the body to defend itself against other hostile organisms. One or two such sera are already in existence and their effect has been studied on an empirical basis, but far more fundamental studies will be necessary before this problem is satisfactorily solved.

An animal can be made to tolerate the graft of another if it is injected with the foreign antigens when newly born or still a foetus. Thus there is some mechanism that can recognize alien cells as those of the self, provided acquaintance has been made at an early enough age—an observation showing similarities to the one describing an imprint of sexual behaviour by an early injection of sex hormones. Maybe we are approaching a biochemical Freudianism that states that some of the most important decisions for later life are already made at a very early age. What happens in the first few hours of an individual's life may have indelible effects on the rest of his existence: a sobering thought indeed.

From the very young we can turn to glance at a problem of senescence. One of the observations correlated with ageing has been a change in body protein. A large amount of the body's protein is collagen, a 2800 Å-long fibrous protein forming part of connective tissue in cartilege and elsewhere. Normally collagen exists as a helix. In the tissues, bundles of collagen are lined up, facing in the same direction. Fibres taken from the old have different tensile strengths from those taken from the young, and this deterioration may be correlated with chemical bonding between bundles of collagen. Ageing may also be expressed in terms of chemical differences of the nucleohistone complexes in the cell nucleus, thus pointing a finger at possible changes in basic cell metabolism as the individual approaches his allotted span of life.

BEHAVIOUR

Three rapidly developing areas of biochemistry may have untold consequences on the human condition: cancer research, neurobiochemistry, and cell differentiation.

A recent series of observations has shown that a particular type of virus can cause cancer in chicken cells. This is the Rous sarcoma virus, peculiar in that it does not have its own protein coat; the latter is provided by a so-called helper virus. If the Rous virus and its helper are given to chickens, they tend to get muscle cancer. The reason seems to be that the virus and its helper disturb the normal function of the cell membranes.

The polyoma virus can cause cancer in mice. The cancerous cells manufacture a new type of protein, although it is not yet certain if this occurs through the incorporation of virus DNA into the cell's control mechanism. Thus there are several possibilities: the cell may incorporate DNA from the virus, acquiring genetic material that may lie dormant for generations. Once triggered by some mechanism, its effect may be to alter cell division and change the properties of the cell membrane. By decreasing or altering the specific stickiness that would normally indicate where the cell should be, it can produce the large, overall effects normally associated with cancer.

We also know that DNA exists outside the cell nucleus, in the mitochondria where production of proteins and enzymes and the all-important cell respiration take place. It is possible that if the wrong genetic material gets into the mitochondrial DNA it may lead to defective mitochondrial synthesis and, in particular, to loss of control over cell respiration. In normal cells aerobic glycolysis is restricted to conditions in which oxygen is present, the so-called Pasteur effect. This is missing in cancer cells, so that glucose is wasted. The idea, originally conceived by Warburg, of regarding cancerous cells as having lost control over their metabolic functions, may be linked with the genetic experiments. Metabolic chaos may be but an expression of a genetic defect which produces loss of control somewhere along the line from DNA control to protein synthesis.

DNA action may also turn out to have a notable influence on processes involving memory. Research is still handicapped by the complexity of this field and therefore the difficulty of interpreting results. There are a number of theories, although not one of them claims to have irrefutable evidence on its side. Memory, nerve function, and thus the whole area of neurobiochemistry, must give explanations depending not only on isolated cells but on the inter-connected nature of the brain and the nervous system. We shall have to talk primarily in terms of networks and pathways, recognizing that the properties of the whole may be significantly different from the characteristics of the individual. There is evidence that even measurements of brain secretions and their changes through the administration of drugs cannot give unambiguous answers, since it is not the absolute amount of materials that matters but their metabolic balance in the body at any given time.

For immediate medical purposes, such arguments may appear trivial. To cure the mentally sick, it is enough to know empirically that certain drugs have a given effect, and it hardly matters how this is achieved at the molecular level. Yet if we admit that pathological mental conditions are no more mystical than similar processes in other parts of the body, we must assume that visible, diagnostic properties are the consequences of underlying bio-chemical changes. Only our present ignorance prevents us from seeing the proper correlations.

All the above depends to a large extent on our knowledge of cell differentiation. This complex area has already been discussed (cf. p. 38) and it is only necessary to emphasize once again its fundamental importance to all biochemical studies. In the very first moments of an individual's potential existence, processes must take place and decisions be made that will have the gravest effects for the rest of its life. If anything goes wrong at this point, all our subsequent efforts will be puny in comparison. And indeed it is at this point that techniques of interference will assume major importance.

COMMUNICATIONS

We have suggested that a collection of facts does not make a science; before we can regard a topic as truly scientific it must contribute to our understanding both of ourselves and the world around us. On this criterion, biochemistry has certainly passed the threshold of science, since to the non-specialist its most liberating influence is the insights it provides into ourselves and our environment.

This is only right and proper, since advances in biochemistry will be mirrored in developments affecting ourselves, our ecology, and our total environment. A curious observation is that the findings of biochemistry seem to confirm in their way older and perhaps humbler views of man.

We can recognize our affinities to far less-developed organisms as well as to our fellow humans, and at the same time see our individuality reaffirmed in our biochemical make-up. The similarity between man and bacterium may not be obviously relevant to a general philosophy, but it does suggest considerations in our increasingly mechanized relationships with other members of the animal kingdom. In an unexpected way, biochemistry also makes us pause when we agree a trifle too readily with cause-and-effect ideas currently put forward on a variety of topics.

Biochemical systems of great complexity cannot be discussed on a teleological basis; we cannot say that the DNA transcription system has evolved because it proved to be most effective for particular ends. Often we are not at all certain why a given mechanism should be present at all. We can say, for example about immune reactions, that their presence is no more than a consequence of innate chemical dispositions. In large areas of biochemistry there seems to be no reason for particular systems. We can only guess that some primeval statistical accident had provided a framework and subsequent developments occurred within it, rather than along revolutionary lines.

In the 1930s the description of elementary particles in physics provided a cautionary tale for cause-and-effect enthusiasts. It can be argued that in our days biochemistry fulfils the same task in showing the possibilities of a statistically based system in operation. This may mean no more than an admission that in our present state of knowledge, when faced with a sufficiently large number of complicated entities, we seek refuge in numbers. It does, however, suggest that there are more ways than one of regarding the advance of a discipline and its implications for the rest of our thinking.

The flexibility such a point of view allows can best be seen in the fact that both general similarity and individuality can find a biochemical home. Although we are all alike biochemically, our own individuality is built into the very fabric of our bodies— perhaps a timely warning for those who would make us all the same. Biochemistry has solved in scientific terms the dilemma that all societies face: the split between the need to conform and the equal necessity of allowing the individual to express his own nature.

There are two broad areas of benefit that biochemistry can bequeath to society. One is the obvious spin-off that any developing discipline offers: improved techniques and their translation into socially acceptable forms. Biochemistry will increase our power over disease, genetic tragedies, the care of the old. It will make significant contributions to our knowledge of thought and motivation. It will illuminate recesses of our nature. It will prove valuable in industry, making possible, or more economical, the production of valuable materials. But the second range of biochemical contributions is just as important; if biochemistry is to have a meaningful ideology, it must be advice about individuality, respect for our environment, the value of diversity in areas of agreement.

There is one major criterion that has to be met before biochemistry can make its best scientific and social contributions. Some of its greatest advances have been made by scientists not trained in biochemistry in any formal sense. Currently, the complexity of biochemical problems more and more demands interdisciplinary teams. Thus the first necessity is for scientists working in the general field of biochemistry to be able to communicate with those in adjoining areas. The history of older sciences shows that at some stage they tend to become inward-looking and exclusive; the language used increases its complexity and becomes foreign even to scientists working in another field of the same discipline. The rapid growth of knowledge is a standing invitation to professional exclusivity, and indeed it is not uncommon to hear a scientist complaining that his own colleagues are beyond understanding.

Such a development would be a disaster for biochemistry. It would stop the interchange with other sciences and tend to develop a specifically biochemical point of view, whereas the strength of biochemistry is its borrowing and synthesis of notions and techniques from other disciplines. Stagnation will only be avoided if a chemist, a biologist, or an engineer, interested in biochemical problems, will still be able to join after formal training in another discipline. This he will only be able to do if he can gain enough biochemical understanding without having to jettison all he has learned and thus render himself useless for the demands of teamwork.

Before condemning such an open-ended policy as reactionary, or tending to dilute the rigour of the discipline, we should be reasonably clear what a biochemist is. The answer is not quite so simple as the question would suggest. Up to a few years ago it was physically impossible for an intelligent schoolchild to decide to become a biochemist through established channels of education, in the same way that he might become an engine driver or a chemist. There were virtually no courses in biochemistry at undergraduate level that would regurgitate him at the appropriate time with the title biochemist indelibly carved on his brain. All biochemistry was taught essentially at the post-graduate level, so that the foundations of some other scientific discipline had first to be acquired. The result was that a number of those who have brought about striking advances in biochemistry have not been biochemists at all, in the sense that a chemist or a physicist is a member of his profession. They were biologists and physicists, physiologists and engineers, chemists and medical men, in the same way that in an earlier period the chemical pioneers may have been physicians or that today's virologist would probably have obtained his training in another discipline. It is an interesting speculation to try to correlate this influx of interdisciplinary scientists with the flowering of biochemistry after the doldrums of biology in the first part of this century. One can also argue that biochemistry is too serious a business to be left to biochemists and can suggest that future advances of the discipline will to a large extent depend on the formation and working together of interdisciplinary teams each member of which can make a contribution to the sum total of inquiry.

Communications within and between disciplines are one factor. Another is communications with those outside the sciences, and here biochemists have a double responsibility. Scientific research is a social phenomenon: society is asked to foot the bill. Although biochemistry is still a cheap science compared, for example, to physics, it is unlikely to remain so for long. Society will, rightfully, be interested in an account of moneys spent and, whatever our own private feelings on the matter, it will simply not be sufficient to argue that research is done because it is interesting to do. It must become painfully evident even to the most blinkered scientist that decisions on the support of research are generally made by people without scientific training. They may be advised by scientists, but the final decision is theirs, arrived at under the influence of the total socio-economic environment. A good decision must be an informed decision, and if biochemistry is to flourish in the future, financial support will be of equal importance to the availability of brainpower. It is up to those who derive their living and enjoyment from biochemistry to make sure that the social decisions that matter are in their favour and that they do not fall into the sin of omission and suffer the results.

This argument can be taken a stage further, through the nature of biochemistry itself. In a number of sciences discoveries appear to be so far removed from the ordinary human condition that they can be shrugged off, at least temporarily; not so in biochemistry. In various areas emotional involvement or mystical fear prove as strong as scientific judgement. This is understandable. Biochemistry may be using rats for experiments and may talk about bacteria, but it takes little imagination to transfer its findings to our own bodies and envisage the possible horrors to come. Apocalyptic imagination is no substitute for scientifically-informed opinion leading to public consensus. It is the task of those who have the well-being of biochemistry at heart to make sure that a general agreement is created and maintained.

FUTURE CONDITIONAL

Conditions for the progress of any science are becoming more complicated and at the same time more challenging. The essence of progress is an abundance of new ideas together with good communications between protagonists. This subject leads directly into a consideration of the explosion, real or imagined, of scientific communications in the last few decades. This topic is outside our scope. But one can summarize current feeling by noting that the computer has been roped in to store, sort, and repackage information, even if its use requires large amounts of money. The problem is not insoluble, it only needs financial support.

Good communications are not much use if there is nothing significant to communicate. After all the mechanical, electronic, and other aids have been discounted we still have to consider the individual scientist sitting in his bath and having an inspired new look at data. Provided there are enough of them, any science will flourish. But brilliant scientists have to be carefully nurtured and the requirement of a happy biochemical future is that they should come from diverse disciplines. Thus biochemistry must offer a sufficient challenge and interest not only to the aspiring undergraduate but also to the fully fledged scientist of another discipline.

This is no easy task. We have already suggested that there is a current disenchantment with science among the young. It is arguable if this is based on reasonable grounds, but the existence of the phenomenon cannot be denied. It is therefore of absolute importance that the ideas and inspirations of biochemistry are projected in such a way that the Fischers, Kendrews, and Khoranas of the future may feel it worth while to undergo the severe training and apprenticeship such a life requires. There are variations in the severity of this problem from one country to another, but in some form it exists in all the civilized world.

Thus another aspect of biochemical progress must be international co-operation both in the field of recruiting the scientists of the future and in correlating large research programmes that may not be viable even on a national basis. By and large, biochemists have been very good on international co-operation. Only recently a European federation of biochemical societies has been established that will perhaps one day result in world-wide cooperation.

Another important group of criteria have little to do with science but control the amount of money available for research. These are political decisions, irrespective of funding.

There are two main sources of money, directly or indirectly—industry and government, and of the two, government is usually by far the larger. Industrial support of biochemistry, especially where it borders on medicine, is bound to grow as its potentials are more clearly realized. But industrial support, as chemists have long realized, is usually and reasonably tied to specific problems whose solution is in the interests of an individual firm. It may prove difficult to persuade those who hold the purse strings that accountable financial support is well advised for unspecified long-term projects.

It is ironic that the best period of growth for any science is war-time, when suddenly a lot of money is available for any activity that might conceivably help the military. We cannot do otherwise but assume that war in its true meaning will not happen again. Yet military or defence spending will increase and so will government support for scientific endeavour that can be described in these terms, regardless of whether, in reality, the particular piece of research has any possible military uses. At the same time, non-defence spending must also show long-term increases, if more slowly, despite temporary cuts brought about by the economists' habit of treating their subject as a science instead of a fallible art.

In all these areas there will arise a new species of research leader, the politician-scientist, who will have to possess not only impeccable scientific credentials but also the ability to press his case in competition with all others clamouring for a share in the available funds. Neither will it do to deride the endeavours of such a new-style scientist, for only through his efforts will it be possible for any discipline to continue to prosper, or even to stay in existence.

Once government or industrial spending on biochemistry reaches economically significant levels, problems of control will present themselves in an acute form. Our society is hardly able to deal with technological innovation even now; it finds itself perplexed whenever a new scientific discovery is presented. There is therefore a strong case to establish some mechanism that would allow scientists, those who convert science into technology and those who make political decisions, to be able, if not to learn each other's jobs, at least to understand what the others are talking about. This is not a plea for the politically-oriented scientist whose scientific excellence is thrown in relief by his political naïveté, but rather for some mechanism whereby men of reasonably good will can come to an agreement before the tide of events renders all discussion futile. Western society had roughly twenty-five years to build a framework for the eventual utilization of atomic power. The chance was missed. It is unlikely that we shall have that much time to prepare for the next round of major scientific discoveries that are more than likely to come from the biochemical side.

Index

ABOUT THE AUTHORS

Peter Farago, currently editor of *Chemistry in Britain,* was educated in Hungary, Switzerland, and England, and received his Ph.D. in chemistry from University College, London. He is a Fellow of Britain's Chemical Society, its Royal Institute of Chemistry, and the British Association of Science Writers, and author of several books on chemistry published in England. He was born in Budapest, Hungary, in 1932.

John R. Lagnado was educated at the University of Geneva and McGill University, where he received his Ph.D. degree in biochemistry. He is a Scientific Fellow of Britain's Zoological Society and is a member of the British Biochemical Society, the International Society of Neurochemistry, and the British Society for Social Responsibility in Science. He has contributed articles to various journals of neurochemistry and biochemistry, and is a member of the Department of Biochemistry, University of London. He was born in Alexandria, Egypt, in 1933.

VINTAGE POLITICAL SCIENCE
AND SOCIAL CRITICISM

VINTAGE BELLES–LETTRES